Do you know as much about science as an average high school science student?

Or a New York stockbroker? Or a housewife in Kenosha, Wisconsin?

More than just another trivia book, this companion volume to the acclaimed PBS series broadcast of "The National Science Test" doesn't just tell you which answers are right and which are wrong, but *why* they are right and wrong. And it provides test scores of groups of people around the country to show you how you measure up.

If you wonder how much you've forgotten since you left school . . . or how much you have to learn to get into college . . . or whether you know more about science than your children, parents, teachers, or friends—find out something fascinating about practically everything, with

The NOVA National Science Test

Ted Bogosian is the producer of "The National Science Test," I and II, as well as such other NOVA broadcasts as "The Cobalt Blues", "Frontiers of Plastic Surgery," and "Tracking the Supertrains." The NOVA television series is a production of WGBH-TV, Boston.

All drawings are by Mark Fisher, for NOVA, except for the drawings on pages 84 and 170, which are by Milton Glazer, for NOVA.

Book design by Paul Souza, WGBH Design

The NOVA television series is produced for PBS by WGBH Boston. Funding for the series is provided by Public Television Stations, Allied Corporation, and the Johnson & Johnson Family of Companies.

Other NOVA books include:

NOVA: Adventures in Science

The NOVA Space Explorer's Guide: Where to Go and What to See

The NOVA National Science Test

By Ted Bogosian and WGBH Boston

A PLUME BOOK
NEW AMERICAN LIBRARY
NEW YORK AND SCARBOROUGH, ONTARIO

NAL books are available at quantity discounts
when used to promote products or services.
For information please write to Premium
Marketing Division, New American Library,
1633 Broadway, New York, New York 10019.

Copyright © 1985
by WGBH Educational Foundation and Ted Bogosian
All rights reserved

PLUME TRADEMARK REG. U.S. PAT. OFF. AND FOREIGN COUNTRIES
REG. TRADEMARK—MARCA REGISTRADA
HECHO EN HARRISONBBURG, VA., U.S.A.

SIGNET, SIGNET CLASSIC, MENTOR, PLUME, ME-
RIDIAN and NAL BOOKS are published *in the United
States* by New American Library, 1633 Broadway,
New York, NY 10019, *in Canada* by The New
American Library of Canada Limited, 81 Mack Av-
enue, Scarborough, Ontario M1L 1M8

Library of Congress Cataloging-in-Publication Data

Bogosian, Ted.
  The NOVA national science test.

  1. Science—Examinations, questions, etc. 2. NOVA
(Television program) 3. Television in science
education. I. WGBH (Television station: Boston, Mass.)
II. Title.
Q182.B64 1985   507'.6   85-13674
ISBN 0-452-25771-9

First Printing, October, 1985

1 2 3 4 5 6 7 8 9

PRINTED IN THE UNITED STATES OF AMERICA

This book was much improved by my friend and colleague, Josephine Patterson, who crafted some of its multiple choice questions and meticulously researched many more. Josie deserves much of the credit for the book's content.

Alain Jehlen and Nancy Linde also deserve praise for their excellent work on the NOVA programs, National Science Test I and II. Without their skill and enthusiasm, neither of the programs would have been as enjoyable and profitable to produce.

Finally, I wish to thank NOVA Executive Producers John Mansfield and Paula Apsell, whose guidance, experience and good humor allowed the purpose of expanding the public interest in science and science education.

Happy Testing!

Theodore Bogosian
Producer
NOVA

This book is dedicated to my parents,
Theodore and Natalie

Contents

Technology	**2**
Life Sciences	**34**
Creatures and Environments	**60**
History and Personality	**102**
Physical Sciences	**150**
Sample Group Test Scores	**194**

The NOVA National Science Test

TECHNOLOGY

Technology

High Speed Trains

New technology has dramatically improved rail service capabilities in recent years.

Today, the introduction of high-speed rail service is being planned between major metropolitan areas in several countries around the world.

Q1 Which of the following is the world's fastest commercial passenger train?

A. The Japanese Shinkansen, also known as the "bullet train"

B. The U.S. Amtrak Metroliner

C. The French TGV or *Train à Grande Vitesse*

D. The West German Magnetic Levitation Train Transrapid

Computer Bits

The Random Access Memory (RAM) silicon chip is quite small, but it is capable of storing a remarkable amount of computer information for use in data processing systems.

Q2 How many bits of information can be stored in the commercially available RAM chip with the largest memory?

A. 1,048,000 bits

B. 64,000 bits

C. 256,000 bits

D. 920 bits

A1

Which of the following is the world's fastest commercial passenger train?

C is the correct answer. The French TGV, which travelled at 236 miles per hour in February 1981, broke the speed record held by the Japanese Shinkansen.

The Amtrak Metroliner achieved its top speed of 162 miles per hour on a New Jersey test track way back in 1963, while the German Magnetic Levitation Train Transrapid has actually travelled faster than the TGV in test runs but has not yet carried commercial passengers.

Japanese magnetic levitation trains (also known as "maglevs"), using a technology much different from the Germans', have also proved promising in recent test runs, exceeding 300 miles per hour over test tracks several miles long.

A2

How many bits of information can be stored in the commercially available RAM chip with the largest memory?

A is the correct answer. The biggest-memory RAM chip can store 1,048,000 bits of information.

Modern computer chips like the RAM are about one-fourth the size of a fingernail and cost about 20 dollars. Each chip has about the same data processing power as Univac 1, one of the first electronic computers, which took up an entire room and cost $701,000 when it was built in 1951.

Technology

Platinum

The precious metal platinum is one of the world's financial standards. Like gold, copper and silver, it is traded at New York commodities exchanges.

But platinum is also at least as valuable within the world's defense establishments, because its heat resistance helps jet fighter engines perform up to exacting military standards.

Q3 What percentage of platinum must the U.S. import to meet its national defense needs?

A. 2 percent
B. 22 percent
C. 60 percent
D. 99 percent

Q4 Instead of silicon, what element is used in semi-conductors for extremely high capacity "supercomputers"?

Q5 What industrial process uses heat and chemicals to extract metals from naturally occurring metallic ores?

A3

What percentage of platinum must the U.S. import to meet its national defense needs?

D is the correct answer. Ninety-nine percent of the platinum metal used for U.S. national defense has to be imported, primarily from two places—South Africa and the Soviet Union.

In case the pipeline of imported minerals should suddenly be disrupted, stockpiles of several years worth of platinum and other so-called "strategic minerals" for which the U.S. is import-dependent—such as cobalt, chromium, and manganese—have been amassed in federal depots throughout the U.S.

Recycling, developing new sources of mineral supplies and substitution for these minerals whenever possible are other means of reducing the U.S.'s vulnerability to occasional disruptions in foreign sources of supply.

A4

Instead of silicon, what element is used in semiconductors for extremely high capacity "supercomputers"?

Gallium

A5

What industrial process uses heat and chemicals to extract metals from naturally occurring metallic ores?

Smelting

Nuclear Power Plants

Nuclear power plants have been strongly criticized in recent years for their potential hazards to human health and safety.

Back in 1970, it was projected that about 300 nuclear power plants would be generating commercial electric power in the U.S. in 1985.

Q6 **How many nuclear power plants are now licensed to generate commercial electric power in the U.S.?**

A. 256
B. 80
C. 161
D. 47

Telescopes

The world's most powerful telescope is sited in the Caucasus Mountains of the USSR. It is 138 feet high, weighs 827 tons and took 16 years to build. Looking through it, one can view the very edge of our universe.

Q7 **The light gathering power of this telescope would enable it to detect the light from a candle at a distance of how many miles?**

A. 15
B. 150
C. 1,500
D. 15,000

A6

How many nuclear power plants are now licensed to generate commercial electric power in the U.S.?

B is the correct answer. In the U.S., construction of approximately 100 nuclear plants has been abandoned after a total investment of more than 9 billion dollars, although 50 plants are still under construction.

Since the March 1979 accident at the Three Mile Island facility outside Harrisburg, Pennsylvania, the safety of nuclear power plants has been of increasing concern to the U.S. public. Equally troublesome has been the combination of skyrocketing cost projections for nuclear reactors and the drop in U.S. power demands.

Current estimates predict that the peak number of nuclear plants in operation in the U.S. by 1990 will be around 120.

A7

The light gathering power of this telescope would enable it to detect the light from a candle at a distance of how many miles?

C is the correct answer. The world's most powerfully ranging of all telescopes is the Soviet Union's altazimuth, mounted 6,830 feet above sea level.

It can locate extraterrestrial objects down to the 25th magnitude. A star observed at this magnitude would appear 100 million times brighter than it would appear to the naked eye.

Technology

Aviation

In 1903, Orville Wright flew at a speed of 30 miles per hour on his first 12-second flight in Kitty Hawk. The *Spirit of St. Louis* averaged 107.5 miles per hour in 1927 during its historic flight from New York to Paris. Today, the Concorde is the fastest flying commercial airplane.

Q8
At what speed does the Concorde cruise as it crosses the Atlantic Ocean?

A. 700 miles per hour

B. 1,350 miles per hour

C. 2,100 miles per hour

D. 3,850 miles per hour

Fiber Optics

Optical fibers act like pipelines for light, and as these fibers may be flexible or rigid, they are proving most useful in a number of fields, especially in medicine and telecommunications. Coded light pulses or complete images are passed through a fiber and analyzed by a detector on the other end.

Q9
What is the most appropriate material for fiber optic production?

A. Glass

B. Copper

C. Steel

D. Plastic

A8

At what speed does the Concorde cruise as it crosses the Atlantic Ocean?

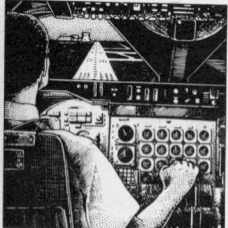

B is the correct answer. The Concorde is a supersonic jet with a Mach 2 air speed potential, meaning it is capable of flying twice the speed of sound. While it flies 1,350 miles per hour over the ocean, the plane slows down as it nears land—to conform with noise pollution regulations—to around 700 miles per hour, approximately the same speed as a jumbo jet.

While "Lucky Lindy" flew solo from New York to Paris in 33 hours and 39 minutes, the passengers and crew of the Air France and British Airways Concorde make the same trip in one-tenth the time—three hours and 45 minutes.

A9

What is the most appropriate material for fiber optic production?

A is the correct answer. Glass, manufactured by special processes and from very pure raw materials, absorbs small amounts of the light passing through it, less so than any other practical material, and so aids the light's ability to retain its brightness and intensity.

Technology

Manhattan Project

The Manhattan Project was the U.S. response to the splitting of the atom in 1939 by two German scientists. European refugee scientists prompted Albert Einstein to alert Roosevelt to what this development could lead to, and soon a group of U.S. and refugee scientists were working secretly at Columbia University in the field of atomic fission.

Q10 What was the outcome of the Manhattan Project?

A. Failure
B. Hydrogen bomb development
C. Nuclear submarine
D. Atomic bomb development

Q11 What is the name of the hydrogen bomb whose primary danger is not heat or a concussive blast but particle radiation?

Q12 What Nazi German rocket was brought to the U.S. by Werner von Braun and developed into intercontinental missile and satellite booster rockets?

A10

What was the outcome of the Manhattan Project?

D is the correct answer. The Manhattan Project had some of the most talented U.S. scientists working together, men like Edward Teller and Robert Oppenheimer. The team was successful in creating the atomic bomb that was dropped on Hiroshima in 1945.

An atomic bomb explodes from a fission reaction, while a hydrogen bomb explodes from a thermonuclear fusion reaction. The H-bomb is 1,000 times more destructive, and it was developed in the 1950s at the insistence of Edward Teller.

The first nuclear-powered submarine, the *Nautilus*, was launched January 21, 1954 after being christened by Mamie Eisenhower, wife of the president.

A11

What is the name of the hydrogen bomb whose primary danger is not heat or a concussive blast but particle radiation?

Neutron Bomb

A12

What Nazi German rocket was brought to the U.S. by Werner von Braun and developed into intercontinental missile and satellite booster rockets?

V-2

Technology

NASA

Since the Soviet Union launched its *Sputnik* satellite back in 1957, humankind has raced to unlock the secrets of space.

In 1977, the U.S.'s National Aeronautical and Space Administration (NASA) launched two unmanned spacecraft to explore two distant planets. One of the spacecraft surveyed the planet Saturn, the other Jupiter and Saturn.

Q13 Which NASA spacecraft surveyed these planets?

A. *Viking I* and *II*
B. *Voyager I* and *II*
C. The *Apollo* Project
D. *Skylab*

Q14 In what year was the first U.S. space shuttle launched?

Q15 What was the first satellite to conduct television signals between the U.S. and Europe?

A13

Which NASA spacecraft surveyed Jupiter and Saturn?

B is the correct answer. *Voyager I* travelled one billion miles in three years before it passed by Saturn n early 1979. *Voyager II* took a longer route, 1.4 billion miles, before it passed by Jupiter and Saturn in 1980–81.

Viking I and *II* were the spacecraft that flew by Mars beginning in 1977. The *Apollo* Project put the first man on the moon in 1969.

Skylab was the first U.S. space station. Launched in 1973, *Skylab* fell back to earth six years later, scattering itself across several thousand miles of the Indian Ocean and Australia.

A14

In what year was the first U.S. space shuttle launched?

1981

A15

What was the first satellite to conduct television signals between the U.S. and Europe?

Telstar

Technology

Fusion

Thermonuclear fusion is the formation of a heavier, more complex nucleus by the joining together of two or more lighter ones. Consequently, large amounts of heat energy are released.

Laboratory experiments have attempted to harness this incredible energy.

Q16 What is the highest temperature yet achieved in a controlled lab setting?

A. 1 million degrees C
  (1,800,000 degrees F)

B. 10 million degrees C
  (18,000,000 degrees F)

C. 82 million degrees C
  (147,600,000 degrees F)

D. 350 million degrees C
  (630,000,000 degrees F)

Microscopes

Faith is a fine invention
When gentlemen can see;
But microscopes are prudent
In an emergency.
—Emily Dickinson, circa 1880.

The world's most powerful microscope has produced photographs of electron clouds of atoms of neon and argon.

Q17 How many times can it magnify something?

A. 260 million times

B. 20 million times

C. 2 million times

D. 2.6 billion times

A16

What is the highest temperature yet achieved in a controlled lab setting?

C is the correct answer. The highest effective laboratory temperature reached was 82 million degrees Celsius at Princeton University's Plasma Physics Laboratory during fusion research in May 1980.

A17

How many times can it magnify something?

A is the correct answer. The electron microscope at the University of Michigan uses lasers to decode holograms produced with KEV radiation, magnifying images up to 260 million times bigger than they really are.

Technology

Concrete

One of the most important materials of modern society is man-made rock: concrete. Concrete is made by mixing cement with water and sand.

Q18 What happens to the water when concrete hardens?

A. It evaporates

B. It becomes part of the concrete molecule

C. The oxygen becomes part of the concrete and the hydrogen is released

D. The hydrogen becomes part of the concrete and the oxygen is released

Architecture

The Roebling Wire Rope Company was founded in the 1830s in Saxonburg, Pennsylvania. It manufactured the first steel wire rope in the U.S. The company bears the name of a man whose most famous design is a graceful national landmark.

Q19 What is the name of that landmark?

A. The Statue of Liberty

B. The Washington Monument

C. The Brooklyn Bridge

D. The Mount Washington Tramway

A18

What happens to the water when concrete hardens?

B is the correct answer. The water becomes chemically and permanently attached to the calcium, aluminum and silicon oxides in the cement. Although concrete has been used for more than 2,000 years, the exact way in which it forms is still being studied.

A19

What is the name of that landmark?

C is the correct answer. John Roebling, a gifted architect, mathematician and engineer, famous for his innovations in the construction of suspension bridges, designed the Brooklyn Bridge in 1866. Three years later, when his plans were accepted, Roebling died of tetanus after a waterfront accident. His son, engineer Washington Augustus Roebling, and Washington's wife Emily took over. They supervised the 14-year project—the end result being what was then the longest suspension bridge ever built. The Brooklyn Bridge spans 1,595 feet across New York's East River, with each of its four main cables containing more than 3,500 miles of wire—enough to stretch across the U.S.

Skyscrapers

Steel frames, reinforced concrete and electric elevators gave birth to the skyscraper—a triumph of U.S. design and construction engineering. Sixteen stories were reached in 1903, but the first true skyscraper came ten years later with 52 stories.

Q20 **What was the name of that building?**

A. The Ingalls Building
B. The Empire State Building
C. The Sears Tower
D. The Woolworth Building

Q21 **What is the longest suspension bridge in the U.S.?**

Q22 **What is the tallest monument in the U.S.?**

A20

What was the name of that building?

D is the correct answer. The Woolworth Building in New York City was designed by Cass Gilbert and was built in 1911. With its Gothic style and 52 stories, it is an important landmark in the history of skyscrapers.

The Ingalls Building was the 16-story building built in Chicago in 1903.

The Empire State Building was inaugurated in May 1931 and is the fourth highest building in the world, standing at 1,250 feet–102 stories.

The Sears Tower is currently the world's tallest building. At 110 stories, it is 100 feet taller than the twin towers of New York's World Trade Center. And it has a population of 16,700 people, 103 elevators, 18 escalators and 16,000 windows!

A21

What is the longest suspension bridge in the U.S.?

Verrazano-Narrows Bridge, New York, NY

A22

What is the tallest monument in the U.S.?

Gateway Arch, St. Louis, MO

21st Century Technology

When this technology was first developed, no one knew what to do with it.

It was first used in surveying. Later, it measured the distance from the earth to the moon to an accuracy of one foot.

Today, its list of uses is endless. It helps print newspapers, cut cloth, remove tatoos, make movies and it may transform conventional warfare.

Q23 It's called the technology of the 21st century. What is it?

A. Fiber optic technology
B. Laser technology
C. Silicon technology
D. Satellite technology

Q24 What kind of satellite has a speed that exactly matches the speed of the earth's rotation?

Q25 What operation separates the components of a liquid mixture?

A23

It's called the technology of the 21st century. What is it?

B is the correct answer. A laser is a device that amplifies the intensity of a light beam by stimulating the energy level of its atoms. Beams of different wavelengths all have different applications ranging from the cutting and welding of metals to attaching retinas to eye tissue.

Fiber optic technology allows light to pass through transparent materials with less absorption than any other practical transmission material.

Silicon technology is a generic term for various industrial applications for the chemical element, including the production of alloys, semiconductors, super- and hype-pure chemical compounds and other high-tech materials.

Satellite technology uses earth-orbiting spacecraft to reflect, receive and/or retransmit electronic signals from one earth-based communications station to another.

A24

What kind of satellite has a speed that exactly matches the speed of the earth's rotation?

Geostationary or synchronous

A25

What operation separates the components of a liquid mixture?

Distillation

Technology

Oil Tankers

The *Seawise Giant* is the world's largest supertanker. It sails under the Liberian flag and ships crude oil around the world.

Q26 If one ton of crude equals approximately 6.8–8.5 barrels of oil, how many tons of oil can this tanker carry?

A. 1,876 dead weight tons

B. 555,843 dead weight tons

C. 400 dead weight tons

D. 1 million dead weight tons

Synthetic Fuels

The impetus to develop the so-called "synthetic fuels"—tar sands, liquified coal and oil shale, among others—has traditionally fluctuated with the price and availability of imported oil.

Underneath the Piceance Basin of Colorado, Utah and Wyoming lie enormous amounts of oil shale, the combination of rock and ancient organic material from which oil can be extracted.

Q27 How many barrels of shale oil are believed to exist beneath the Piceance Basin?

A. 6 billion   C. 600 billion

B. 60 billion   D. 6 trillion

A26

If one ton of crude equals approximately 6.8–8.5 barrels of oil, how many tons of oil can this tanker carry?

B is the correct answer. Depending on the temperature of the oil and its specific gravity, 555,843 dead weight tons equal between 3,779,132.4 and 4,724,665.5 barrels of oil.

A27

How many barrels of shale oil are believed to exist beneath the Piceance Basin?

C is the correct answer. Six-hundred billion barrels of shale oil are thought to exist beneath the Piceance Basin. This is approximately enough to satisfy present U.S. oil demands until the 25th century.

While the technology to extract oil from shale has existed for almost a century, economic and environmental problems have kept U.S. shale oil projects at the pilot stage pending drastic rises in crude oil prices above 1979 levels.

Technology

Acronyms

SONAR is an acronym derived from the first two letters of the words **SO**und and **NA**vigation and the first letter of **R**anging. It is a branch of applied acoustics concerned with utilizing fluids as transmitting media.

Q28 Which of the following is not an acronym?

A. ATOM
B. AIDS
C. PABA
D. LASER

Q29 In medical imaging, what do the initials CAT and NMR represent?

Q30 What is the name for the Navajos' round, earth-covered dwelling?

A28

Which of the following is not an acronym?

A is the correct answer. The word atom derives from the Greek *atomos* meaning indivisible.

AIDS, is medical shorthand for **A**cquired **I**mmunological **D**isease **S**yndrome or in the military, for **A**ircraft **I**ntrusion **D**etection **S**ystem.

PABA stands for **PA**ra-amino**B**enzoic **A**cid, a compound often found in suntan lotion.

Laser stands for **L**ight **A**mplification by **S**timulated **E**mission of **R**adiation. It is the coherent light beam produced by exciting the atoms of certain materials.

A29

In medical imaging, what do the initials CAT and NMR represent?

Computerized Axial Tomography and Nuclear Magnetic Resonance

A30

What is the name for the Navajos' round, earth-covered dwelling?

Hogan

Technology

Solar Energy

Socrates had this to say about building homes:

"Now in houses with a south aspect, the sun's rays penetrate into the porticoes in winter, but in summer the path of the sun is right over our heads and above the roof, so that there is shade. If then, this is the best arrangement, we should build the south side loftier to get the winter sun and the north side lower to keep out the cold winds."

Q31 **This reasoning represents the use of what type of solar energy?**

A. Active
B. Biomass
C. Passive
D. Photovoltaics

Units of Measurement

Like most people, scientists often live, work and die in relative obscurity, but some scientists may enjoy a bit of immortality by having something named after them, like a unit of measurement.

Q32 **Which of the following units was *not* named for its discoverer?**

A. Tesla
B. Farad
C. Mach
D. Erg

A31

This reasoning represents the use of what type of solar energy?

C is the correct answer. A design that manages to take advantage of the sun with few or no moving parts is called "passive" solar energy.

Active is just the opposite and may use one or a variety of mechanisms to collect and store solar energy.

Biomass includes all plant matter that is used to produce energy, not just wood, but also technologies to improve yields of sea algae farms, research on gasahol and gasification of manure.

Photovoltaics is an active technology that falls in the category of solar electrics. Photovoltaics converts light to electricity by using semiconductor technology, the basis of the transistor and integrated circuit industry.

A32

Which of the following units was *not* named for its discoverer?

D is the correct answer. An erg is a unit of work or energy: the amount of work done when a force of one dyne moves its point of application one centimeter in the direction of force.

A tesla is a unit of magnetic flux density named after inventor Nikola Tesla.

A farad, named after 19th century English physicist Michael Faraday, measures transmission of electrical energy.

A mach, or mach number, is the ratio of the speed of a body to the speed of sound in the surrounding atmosphere, named after 20th century Austrian physicist Ernst Mach.

Technology

Agent Orange

Agent Orange is the common name for a 50%–50% mixture of the herbicides 2,4,5–T and 2,4–D.

During the Vietnam War, the U.S. sprayed over 11 million gallons of Agent Orange onto the Vietnamese countryside in an effort to defoliate the thick ground cover that protected enemy troops.

Q33 **What chemical does Agent Orange contain that is considered harmful to life?**

A. PCB
B. Dioxin
C. Sulfur monochloride
D. Halogenobenzenes

Q34 **What kind of energy is extracted from naturally occurring steam and hot water found in the earth?**

Q35 **In 1978, the *Double Eagle II* gained aeronautical immortality. Why?**

Q36 **What fibrous, inflammable material used in industrial materials is composed primarily of magnesium and calcium silicate?**

A33

What chemical does Agent Orange contain that is considered harmful to life?

B is the correct answer. Dioxin, also called dimethoxane, is a toxic hydrocarbon best known as an impurity in the herbicide 2,4,5-T. It is one of the most toxic materials known and has a half-life in the soil of about one year. Less than 50-millionths of a gram of dioxin will kill a mouse.

A34

What kind of energy is extracted from naturally occurring steam and hot water found in the earth?

Geothermal

A35

In 1978, the *Double Eagle II* gained aeronautical immortality. Why?

First manned balloon to cross the Atlantic Ocean

A36

What fibrous, inflammable material used in industrial materials is composed primarily of magnesium and calcium silicate?

Asbestos

Technology

Artificial Intelligence

From evil robots to futuristic machines like R2-D2 and C-3PO, movies have created machines that can think like people.

Away from Hollywood, in research laboratories, computer scientists are actually trying to write programs that can imitate human thought. But progress is slow because so little is known about how the human brain thinks.

Q37 **Which of the following human activities has proved the most difficult for computers to mimic?**

A. Medical diagnosis

B. Chess playing

C. Translating poems from one language to another

D. Determining the structure of complex chemicals

Nukes

The first self-sustaining nuclear chain reaction took place in the U.S. The first nuclear bomb was built here as well as the first commercial nuclear reactor.

But now other countries are far more dependent on nuclear power than we are.

Q38 **Which country gets the biggest share of its electric power from nuclear reactors?**

A. France

B. USSR

C. Sweden

D. Japan

A37

Which of the following human activities has proved the most difficult for computers to mimic?

C is the correct answer. Effective computer programs have been written for chess playing and chemical analysis. Even medical diagnosis programs have been fairly successful.

But to make a computer understand the complexities of language is far more difficult. Scientists are still a long way from understanding human thought and communication patterns, and it is this knowledge that is needed before computers can be programmed to think like us.

A38

Which country gets the biggest share of its electric power from nuclear reactors?

A is the right answer. France gets 25 percent of its power from nuclear reactors. Japan gets 14 percent, Sweden 23 percent and the USSR only 5 percent. The U.S. is the world's number one producer of nuclear power, but of all energy consumed in the U.S., only 12 percent is nuclear powered.

LIFE SCIENCES

Life Sciences

Synthetic Insulin

There are hopes that new recombinant DNA techniques will turn out to be a cheap and effective way to make a host of vital products, ranging from fertilizer to medicine. One of the first useful products beginning to be made with recombinant DNA is human insulin for diabetics.

Q39 Using the recombinant DNA process, how would human insulin be created?

A. By mutating a pig's gene
B. By making a gene from common chemicals
C. By putting a human gene into bacteria
D. By splicing together two bacterial genes

Enzymes

An enzyme is a protein generated by a living cell. Enzymes function as catalysts in reactions involving the metabolism of living organisms and so play a vital role in food production, processing and consumption.

Found in the pancreas of many animals, including man, chymotrypsin is a protein-digesting enzyme.

Q40 If a certain protein takes one hour to digest in the presence of chymotrypsin, how long would similar digestion take without it?

A. 5 hours
B. 1 month
C. 100 years
D. 1,000 years

A39

Using the recombinant DNA process, how would human insulin be created?

C is the answer. Recombinant DNA involves taking genes from different species and implanting them in rapidly multiplying bacteria. The genes are then duplicated every time the bacteria reproduce, and large quantities of valuable biochemicals can be made.

A40

If a certain protein takes one hour to digest in the presence of chymotrypsin, how long would similar digestion take without it?

D is the correct answer. At least 900 organic enzymes are known to exist, and their extreme importance to the functioning of the human body is inestimable.

Immunology

Immunology describes the way the human body defends itself against virus attacks.

Once inside the cells of their host, invading viruses use the host cells' own machinery to reproduce themselves. After the host cell dies, the new virus particles swarm out of the dead cell to infect its neighbors.

But the first cell didn't die in vain, because as the virus multiplies, it triggers an alarm in the infected cell's nucleus.

The dying cell responds by manufacturing a virus-blocking substance. This substance then streams out of the dying cell and encircles uninfected cells nearby, creating the first line of defense against further virus infection.

Q41 What is this virus-blocking substance called?

A. Antigen
B. Interferon
C. Lymphocyte
D. Monoclonal antibody

Q42 How long can a parasite live within the human body?

Q43 What science is the study of hormones?

A41

What is this virus-blocking substance called?

B is the right answer. Interferon is produced within hours of the onset of viral infection, and it triggers all of the human body's natural defense mechanisms. It was discovered by a British scientist in 1957, and for a while it looked like a wonder drug. Some doctors thought it could cure everything from colds to cancer.

But in many cases interferon hasn't worked. Researchers hope to manufacture large quantities of synthetic interferon soon, so they can do more testing and find out how effective this natural disease fighter can be.

Antigens are those substances which invade the human body and stimulate the production of antibodies. Lymphocytes are one kind of white blood cell, which the body also manufactures to fight infection, and monoclonal antibodies are some of the body's specific anti-viral agents.

A42

How long can a parasite live within the human body?

15 years

A43

What science is the study of hormones?

Endocrinology

Life Sciences

Cell Energy

All animals and plants are made up of cells.

While larger plants and animals are created by gluing single cells together into larger structures, some lower forms of life, like bacteria, may be nothing more than a single cell.

The major activity of all cells is to produce energy.

Q44 In what part of an animal cell is energy produced?

A. Nucleus
B. Mitochondrion
C. Endoplasmic reticulum
D. Golgi complex

Capillaries

The human body is laced with blood channels of various sizes intricately arranged so that no cell is very far from a channel and the blood it carries.

The widest blood channels are the body's arteries and veins, then come its blood vessels, and its narrowest channels are the capillaries.

Q45 How many miles of capillaries exist in each adult?

A. 10 miles
B. 400 miles
C. 12,000 miles
D. 62,000 miles

A44

In what part of a cell is energy produced?

B is the correct answer. Mitochondria are the organelles in which an animal's cellular energy is produced.

The nucleus contains the cell's genetic material, the endoplasmic reticulum is where proteins are manufactured, and the Golgi complex is the section of the cell that functions in secretion.

A45

How many miles of capillaries exist in each adult?

D is the correct answer. There are about 62,000 miles of capillaries in each adult human, enough to stretch ten times back and forth across the continental U.S. if they were put end to end.

Life Sciences

Cell Division

After a human egg cell is fertilized, the following 24 hours are spent in rest, as if the cell is gathering strength for the difficult job ahead. Then it begins to divide rapidly into new cells.

This cell division represents the human reproduction system.

Q46 What is this cell division called?

A. Meiosis
B. Mitosis
C. Necrosis
D. Catalysis

Q47 To the nearest 100, how many separate bones does the human skeleton contain?

Q48 Name the outer, non-sensitive layer of human skin.

A46

What is it called

A is the correct answer. For several days all the new cells are exactly alike, but by the 12th day two distinct layers of cells have formed. As different cell layers develop, they begin to form the groundwork of a rudimentary human body.

Mitosis is the process by which less evolved animal and plant cells divide; necrosis is cell death, and catalysis is the process by which chemical reactions are facilitated.

A47

To the nearest 100, how many separate bones does the human skeleton contain?

200 (206 bones)

A48

Name the outer, non-sensitive layer of human skin.

Epidermis

Life Sciences

Life Expectancy

The fastest growing age group in the U.S. is the group 75 years of age and older.

By early 1984, babies born in the U.S. could be expected to live an average of 74.5 years.

Q49 What was the life expectancy of people in the U.S. born at the turn of this century?

A. 47 years
B. 50 years
C. 59 years
D. 63 years

Q50 What is the medical name for the condition commonly called "hardening of the arteries"?

Q51 What is unnatural opacity on the lens of the eye called?

A49

What was the life expectancy of people in the U.S. born at the turn of this century?

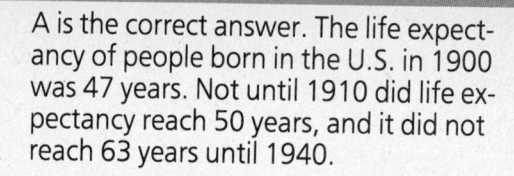

A is the correct answer. The life expectancy of people born in the U.S. in 1900 was 47 years. Not until 1910 did life expectancy reach 50 years, and it did not reach 63 years until 1940.

A50

What is the medical name for the condition commonly called "hardening of the arteries"?

Arteriosclerosis

A51

What is unnatural opacity on the lens of the eye called?

Cataracts

Life Sciences

Epidemics

By the end of the 6th century A.D., approximately half the population of the Byzantine Empire had been killed by the bubonic plague. Eurasia in the 1340s suffered 25 million deaths during the "Black Death," the second worst epidemic to strike humankind.

Q52 In the early 20th century, in history's third worst disease disaster, more than 20 million people died from what illness?

A. Cholera
B. Bubonic plague
C. Influenza
D. Typhus

Q53 What phase of an infectious disease occurs between infection and the appearance of symptoms?

Q54 In 1984, French and U.S. scientists independently discovered the virus that appeared to be the cause of AIDS. What is that virus called?

A52

In the early 20th century, in history's third worst disease disaster, more than 20 million people died from what illness?

C is the correct answer. In the single year of 1918, 21 million people died from influenza, 8 million more than the total number of military casualties in World War I. In the U.S. alone more than one-half million people died, and it has been said that had the influenza epidemic continued its mathematical rate of acceleration, the world's human population would have been eradicated within just a few weeks.

Influenza is caused by a filterable virus that was not isolated until 1933. Since then, vaccines have been developed and many strains of the disease have been discovered within the two main types, A & B.

The 20th century has seen epidemics of other diseases: in 1947, cholera swept through Egypt killing 10,276 people. Two million of India's population died in the 1920s from the bubonic plague, and typhus hit Russia in 1917–21, killing three million.

A53

What phase of an infectious disease occurs between infection and the appearance of symptoms?

Incubation

A54

In 1984, French and U.S. scientists independently discovered the virus that appeared to be the cause of AIDS. What is that virus called?

HTLV-III (US) or LAV (French)

Life Sciences

Psychological Tests

Sigmund Freud summed up his work when he stated, "My life and work has been aimed at one goal only: to infer or guess how the mental apparatus is constructed and what forces interplay and counteract in it."

In working toward that goal, Freud was a founding father of modern psychiatry. Many doctors of this century have worked on developing tests that would expose just how a particular individual's mind works.

Q55 Which of the following tests uses a specially prepared series of inkblots to study a person's perceptual process?

A. The Thematic Apperception Test (TAT)
B. The Rorschach technique
C. The Blacky Pictures
D. The Stanford Binet Scales

Brain

The brain—an information processing system *extraordinaire!*

The brain is a human being's smartest organ and one of its hungriest. The 100 billion nerve cells that make up the brain require 400 calories of fuel per day as well as 20 percent of the body's total oxygen.

Q56 How much does an adult human brain weigh?

A. 8 ounces
B. 1.5 pounds
C. 5 pounds
D. 2.5 pounds

A55

Which of the following tests uses a specially prepared series of inkblots to study a person's perceptual process?

B is the correct answer. Developed by Hermann Rorschach, Swiss psychiatrist, 1884–1922, the Rorschach technique is a psychological test consisting of 10 inkblots of varying designs and colors that are shown to a subject one at a time with the request to interpret them. The purpose is to furnish a description of the dynamic forces of personality through an analysis of the subject's interpretations. This test serves as an aid in problems of differential psychiatric diagnosis.

The TAT is a series of black and white pictures, drawings or woodcuts—most of which portray one or more people in situations designed to elicit themes of psychological significance. The subject is asked to make up a story for each picture and to specify what led to the situation, what is happening, what the characters are thinking and feeling, and how the story ends.

Similarly, the Blacky Pictures are a series of 13 cartoons featuring a dog named Blacky in situations designed to evoke stories and themes relevant to the purpose of the test.

The Stanford Binet Scales is the original IQ test developed in 1904. It was improved upon and today the Wechsler Adult Intelligence Scale is the major adult intelligence test in clinical use.

A56

How much does an adult human brain weigh?

D is the correct answer.

Life Sciences

Endorphine

Some of the drugs now being discovered are manufactured in our own brains. Upon the discovery of the naturally occurring endorphine, scientists sought ways to manufacture this substance synthetically. Why?

Q57 **What function does endorphine perform?**

A. Stimulates fertility
B. Inhibits pain
C. Unblocks arteries
D. Antihistamine

Q58 **What do psychologists call a person who excels in one area while being considerably below average in others?**

Q59 **What term describes two organisms of different species living intimately together?**

Q60 **What is the name of the group of chemicals—including the sex hormone testosterone—used by athletes to increase strength?**

A57

What function does endorphine perform?

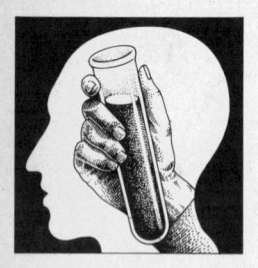

B is the correct answer. Endorphine is a compound similar to morphine. When released in larger than normal amounts it performs a painkilling function and may thus play a role in mood, learning, memory and behavior.

A58

What do psychologists call a person who excels in one area while being considerably below average in others?

Idiot Savant

A59

What term describes two organisms of different species living intimately together?

Symbiosis

A60

What is the name of the group of chemicals—including the sex hormone testosterone—used by athletes to increase strength?

Anabolic Steroids

Life Sciences

Red Blood Cells

Red blood cells travel through tiny blood vessels, carrying oxygen to the body's other cells.

Q61 How many red blood cells are produced by the human body each day?

A. 200,000
B. 2 million
C. 2 billion
D. 200 billion

Q62 What is the liquid component of blood?

Q63 What is the name for a balloon-shaped dilation in a tube?

Q64 What are cancer-causing substances called?

A61

How many red blood cells are produced by the human body each day?

D is the correct answer. Each day, 200 billion red blood cells are produced. They live three to four months, making 170,000 trips around the body. An equal number of red blood cells die each day and are disposed of by white blood cells or leukocytes.

If all the red blood cells in a human body were lined up end to end, they'd stretch all the way around the world.

A62

What is the liquid component of blood?

Plasma

A63

What is the name for a balloon-shaped dilation in a tube?

Aneurism

A64

What are cancer-causing substances called?

Carcinogens

Life Sciences

First Artificial Heart

In 1982, Seattle dentist Barney Clark was given an artificial heart to replace his own, which was failing. The operation was widely hailed as an historic first. The heart kept him alive for 112 days. But Clark was not the first person to receive an artificial heart. Back in 1969, Haskell Karp, a printer's clerk from Skokie, Illinois, was given an artificial heart temporarily, while waiting for a heart transplant.

Q65 Who was the surgeon who performed the operation?

A. Dr. Adrian Kantrowitz
B. Dr. Denton Cooley
C. Dr. Michael DeBakey
D. Dr. Robert Jarvik

Q66 What is the name of the four-square diagram that scientists use to compare dominant and recessive genetic characteristics?

Q67 What is medicine called when it is used in criminal and legal circumstances?

A65

Who was the surgeon who performed the operation?

B is the correct answer. Dr. Denton Cooley of Houston. Although Cooley used an artificial heart that could not pump blood without damaging the blood cells (ones that could had not yet been invented), he hoped it could keep Karp alive long enough for him to find a human heart donor. A donor was found, and three days after the first operation, Cooley replaced the mechanical heart with a human organ. But 31 hours after the second operation, Haskell Karp died.

A66

What is the name of the four-square diagram that scientists use to compare dominant and recessive genetic characteristics?

Punnett Square

A67

What is medicine called when it is used in criminal and legal circumstances?

Forensic Medicine

Oocytes

An oocyte is an unfertilized human egg. Oocytes can now be surgically removed from a woman's ovarian follicle as part of a revolutionary medical technique called in-vitro fertilization.

For infertile couples, this process keeps the egg alive in a Petri dish until it can be fertilized with sperm.

After the fertilized egg begins to develop, another surgical procedure transfers the egg back into the woman's uterus where it continues to develop into a fetus.

This new method actually fertilizes an egg outside the mother's womb.

Q68 **What is the success rate of conception through in-vitro fertilization?**

A. 1 in 5
B. 1 in 50
C. 1 in 200
D. 1 in 500

Human Embryo

In the development of a human embryo, there is at first a cell division of the fertilized egg. The cells form a dense cluster, but by the end of five days, they have arranged themselves into a sphere filled with fluid, called a blastocyst, which implants itself in the mother's uterus. There, the cells specialize and start building body organs.

Q69 **At what age is the developing embryo called a fetus?**

A. 7 weeks
B. 12 weeks
C. 10 weeks
D. 15 weeks

A68

What is the success rate of conception through in-vitro fertilization?

A is the correct answer. Medical progress has made the probability of conception through in-vitro fertilization about the same as through intercourse, one in five. But, not all couples are candidates for in-vitro fertilization. There are strict age requirements, the procedure is expensive and it's only performed in a few hospitals around the world.

A69

At what age is the developing embryo called a fetus?

C is the correct answer. An embryo is called a fetus at about 10 weeks old. It is just two inches long and moves actively.

At 12 weeks, it's three inches long.

By 15 weeks, the sensory organs are almost completely formed.

At 18 weeks, the fetus is five inches long. But it will be at least eight more weeks before it has any chance of living outside its mother's womb.

Calories

Obesity is a common nutritional problem among adults in the U.S. Since a calorie is the energy value of food, extra weight is gained when a person consumes more calories than he or she burns. That extra weight may strain other parts of the body and divert blood from vital organs, one of which is the brain, which ounce for ounce requires proportionately more fuel than any other organ in the body.

Q70 Which of the following nutrients provides the chief source of energy for the brain?

A. Fats
B. Carbohydrates
C. Proteins
D. Vitamins

Q71 What tube carries food from the mouth to the stomach?

Q72 What theory suggests that each body has a predetermined amount of fat, regulated by appetite and caloric expenditure?

A70

Which of the following nutrients provides the chief source of energy for the brain?

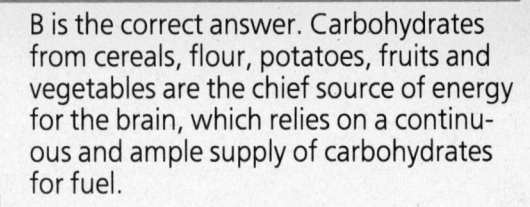

B is the correct answer. Carbohydrates from cereals, flour, potatoes, fruits and vegetables are the chief source of energy for the brain, which relies on a continuous and ample supply of carbohydrates for fuel.

A71

What tube carries food from the mouth to the stomach?

Esophagus

A72

What theory suggests that each body has a predetermined amount of fat, regulated by appetite and caloric expenditure?

Setpoint

CREATURES AND ENVIRONMENTS

Endangered Animals

The National Endangered Species Act was passed by the U.S. Congress in 1973. At that time, approximately 400 species of plants and animals—international as well as domestic—were placed on the list of those approaching extinction, and measures were taken to protect them.

Q73 **Which of the following animals are no longer on the U.S. Department of Interior's Endangered Species list?**

A. Florida manatee

B. African leopard

C. The whooping crane

D. Bald eagle

Giant Tortoises

Giant land tortoises can be found on the Galapagos Islands off South America and on the Seychelles Islands north of Madagascar.

Q74 **How old can these giant tortoises get?**

A. 25 years C. 175 years

B. 75 years D. 300 years

A73

Which of the following animals are no longer on the U.S. Department of Interior's Endangered Species List?

B is the correct answer. Some African leopards have been changed from an endangered species to a threatened species since 1973. This change means that leopards found in or south of Gabon, Congo, Zaire, Uganda and Kenya are farther away from extinction than they were a decade ago, although they are still not out of danger of becoming extinct.

Florida manatees, bald eagles and whooping cranes all remain on the U.S. Interior Department's list of endangered species. There are now around 800 species of plants and animals on the endangered species list, twice as many as a decade ago.

A74

How old can these giant tortoises get?

C is the correct answer. It's very hard to keep track of an animal for that long, of course. But one tortoise actually lived in captivity for 152 years and it was at least 25 years old when it was caught. Several giant tortoises have lived as pets for more than a century, and it is believed that the giant tortoise lives longer than any other animal on earth.

Biochemical Equation

One of the most important of all biological processes is photosynthesis.

From a few simple inorganic compounds and from the sugar made during photosynthesis, come all of the complex kinds of molecules essential to the structure and vitality of life on earth.

Q75 Which of the following is not a part of the biochemical equation for photosynthesis?

A. Nitrogen
B. Carbon dioxide
C. Oxygen
D. Water

Oldest Living Tree

The oldest known living tree is the bristlecone pine tree. Nicknamed "The Patriarch," it's 4,600 years old and still growing strong.

Q76 But where?

A. Jerusalem
B. China
C. U.S.
D. India

A75

Which of the following is not a part of the biochemical equation for photosynthesis?

A is the correct answer. Nitrogen is not a part of the photosynthetic equation. The chemical equation for the reaction is:

$$6CO_2 + 6H_2O \longrightarrow C_6H_{12}O_6 + 6O_2$$

Photosynthesis is a process that converts carbon dioxide and water into sugar. A by-product of this reaction is oxygen, which is essential for animal life. The animals, in turn, produce carbon dioxide when they breathe, which is one of the ingredients plants use to make glucose via photosynthesis.

With minor exceptions, the existence of the entire biological world hinges upon this process.

A76

Where does "The Patriarch" grow?

C is the correct answer. Bristlecone pine trees grow in the eastern Sierra area of California and the White Mountains of Nevada. "The Patriarch" is a scruffy little thing, no more than 40 feet tall. It is not majestic like the giant sequoias, but it is poetic looking in a gnarled way. There was no way of knowing its age until the advent of microscopes, which made it possible to count its many rings.

On summits where these trees often grow, snow may remain for several summers, hindering growth. Winds blowing sand particles from the desert below tear at the trees' bark and branches, and the lack of rainfall causes these trees' slow growth—as well as the microscopically narrow tree rings. As many as 1,100 rings may exist in the space of five inches.

Heaviest Mammal

Five to six feet in length and weighing around 85 pounds, the harbor porpoise is one of the smallest whales. The Great Blue Whale is 3,500 times heavier than this porpoise. It is considered to be the largest animal to have ever lived on the earth.

Generally, these giants migrate annually from their Antarctic feeding grounds to bodies of water near the equator. Fasting every winter, they spend their summers eating.

Q77 How much food do Blue Whales consume every day during their eating period?

A. 200 pounds of krill
B. 1,000 pounds of krill
C. 2.5 tons of krill
D. 4 tons of krill

Q78 What is the practice of cultivating fresh- and salt-water organisms called?

Q79 True or false: Sharks are able to detect and use electric fields in ocean water as a method of navigation.

A77

How much food do Blue Whales consume every day during their eating period?

D is the correct answer. The whales are so huge that they require at least 1.5 million calories per year; krill are very small shrimp-like animals, so that's 40 million krill per whale per day. During the summer months they must have four meals daily, because their stomachs can only hold one ton of food.

Blue Whales have been measured up to 100 feet in length, and the heaviest weighed was 136 tons... more than any dinosaur yet unearthed!

A78

What is the practice of cultivating fresh- and salt-water organisms called?

Aquaculture

A79

True or false: Sharks are able to detect and use electric fields in ocean water as a method of navigation.

True

Creatures and Environments

Bird Egg Size

The egg is the largest single cell in the animal world.

Q80 **Which of the following birds produces the largest egg?**

A. The North African ostrich
B. The kiwi
C. The wandering albatross
D. The bee hummingbird

Mating Habits

Most mammals have a distinct mating or breeding season, as well as courtship behavior peculiar to their species. The male whooping crane, for example, is famous for the beautiful dance it performs in an effort to woo his mate.

Q81 **Which of the following animals is known to eat its spouse after— and sometimes during—the mating process?**

A. The blue crab
B. The scorpion
C. The lightning bug
D. Bumblebee

A80

Which of the following birds produces the largest egg?

A is the correct answer. An average-sized ostrich egg weighs around three pounds 11 ounces with a shell that can support the weight of a 252-pound man.

The kiwi produces an egg that is the biggest in proportion to its size. A kiwi is the size of a hen and puts out an egg 10 times the size of a chicken egg.

The albatross produces the largest egg of any sea bird, one that weighs around 20 ounces and takes quite long to incubate—around 75 days.

The bee hummingbird is the world's smallest bird and produces the smallest of eggs. A specimen at the U.S. National Museum of Natural History weighs only 0.176 ounces.

A81

Which of the following animals is known to eat its spouse after—and sometimes during—the mating process?

B is the correct answer. And scorpions share this trait with members of the Arachnida family, like black widow spiders (hence the name) who eat their mates after their eggs have been fertilized. The preying instinct drives them to do so.

The lightning bug and the bumblebee do not exact such a high price from their mates, and the female blue crab is entertained and sheltered by the male during their courtship and mating periods.

Trees

One species of trees dates back to the time of the dinosaurs. Also called the maidenhair tree, it has wedged-shaped leaves and yellow flowers, is native to China and Japan, but also grows in New York City.

Q82 What is the name of this tree?

A. The weeping willow
B. The sequoia tree
C. The horse chestnut tree
D. The ginkgo tree

The Sahara

The Sahara Desert was not really a desert 5,000 years ago. The area was dotted with large shallow lakes and covered with vegetation. Today, the Sahara is the world's largest and most forbidding desert. The summer temperature can exceed 130 degrees Fahrenheit (54 degrees Celsius) in the shade, and the whole region receives less than five inches of rainfall per year.

Q83 How big is the Sahara Desert?

A. 500,000 square miles
B. 1 million square miles
C. 3.5 million square miles
D. 8 million square miles

A82

What is the name of this tree?

D is the correct answer. The ginkgo dates back to the Paleozoic era. Called a living fossil, it does not appear in the wild, but readily survives cold temperatures and urban pollution.

The willow is younger in origin than both the ginkgo and horse chestnut and does not have wedged-shaped leaves.

The sequoia, a North American native, produces some of the tallest living trees—some known to exceed 275 feet. It should not be confused with the coastal redwood, which is a different species.

The horse chestnut has survived since the Tertiary period, but is native to the Balkans.

A83

How big is the Sahara Desert?

C is the correct answer—or was the correct answer yesterday, anyway, for the earth's desert areas are advancing rapidly. Much of the Sahara Desert was once productive land, but over the centuries, population pressures and inappropriate cultivation and grazing methods have served to increase the size of this formidable desert. Just on the Sahara's southern edge alone, over 250,000 productive acres have been lost.

Tidal Power

The world's largest tidal generation plant, in Brittany, France, draws its power from the enormous amount of sea water that is set in motion by incoming and outgoing tides.

But it is in North America where a 50-foot tidal range, the greatest in the world, is found.

Q84 What is the name of this body of water?

A. Prudhoe Bay
B. Bay of Fundy
C. San Francisco Bay
D. Nantucket Sound

Q85 What process destroys water supplies through excessive algae growth?

Q86 At the mouth of what river is the ecologically unique Atchafalaya Swamp located?

A84

What is the name of this body of water?

B is the correct answer. In the Bay of Fundy, which lies between the Canadian provinces of New Brunswick and Nova Scotia, the tidal range is about 50 feet during the peak spring season.

High and low tides result from the gravitational pull of the moon and the sun upon the earth, and a mean tidal range of 16–17 feet is considered mandatory for a tidal energy plant to be commercially viable. Besides the Bay of Fundy, other North American nearshore locations with the potential for tidal power development include the coasts of Alaska, British Columbia, California and Maine. Internationally, the potential exists on almost every continent. France and South Korea own the most active tidal power projects at present.

A85

What process destroys water supplies through excessive algae growth?

Eutrophication

A86

At the mouth of what river is the ecologically unique Atchafalaya Swamp located?

Mississippi

Creatures and Environments

Spiders

Spiders have been around for 300 million years. While 30,000 different species of spiders have already been identified, more than 100,000 species have yet to be identified by etymologists. Many people are afraid of spiders, but few are actually dangerous to humans.

Q87 However, one spider indigenous to the U.S. has a bite that is sometimes fatal to humans. Which one is it?

A. The wolf spider
B. The tarantula
C. The recluse spider
D. The scorpion

Animal Migration

Throughout history animals and people alike have trekked thousands of miles during pilgrimages and migrations—overcoming harsh environments and the vicissitudes of travel.

But the migration efforts of many animals, which happen year after year, are a constant wonder.

Q88 Of the following animals, which has the longest annual migration?

A. The monarch butterfly
B. The caribou
C. The bowhead whale
D. The wolf

A87

One spider indigenous to the U.S. has a bite that is often fatal to humans. Which one is it?

C is the correct answer. The tiny recluse spider, common in the midwestern and southern states, injects venom from its fangs that is deadlier than the poison from a rattlesnake or black widow spider. The bite may cause gaping and long-lasting flesh wounds, posing the greatest danger to infants and the elderly, who are most vulnerable to its poison.

The U.S. tarantula's painful bite is not life-threatening.

The wolf spider's bite is not harmful to humans.

And the scorpion is not a spider. Scorpion bites are usually not fatal, though tropical species are more dangerous than other ones.

A88

Of the following animals, which has the longest annual migration?

A is the correct answer. During the summer, most monarch butterflies breed in the Great Lakes Region of North America. When autumn comes, they begin their 1,800 mile migration to Mexico. Occasionally, monarch butterflies have been known to cross the Atlantic Ocean. Most don't make it—but some hardy individuals do survive that incredible 5,000 mile journey.

Marsupials

Marsupials are mammals with a unique method of reproduction.

The gestation period for marsupial embryos begins in the uterus but ends in a pouch on the female's abdomen.

Q89 Which of the following is not a marsupial?

A. Wombat
B. Bandicoot
C. Opossum
D. Marmot

Q90 Bears are found on every continent except two. Which two?

Q91 What is the name for the annual period of sexual interest among deer and elk?

A89

Which of the following is not a marsupial?

D is the correct answer. A marmot is a rodent and not a marsupial.

Like their more famous cousin—the kangaroo—wombats, bandicoots and opossums are all members of the marsupial family. Marsupials primarily live in Australia and South America and vary widely in size.

While other mammals' gestation periods end as soon as the fetus is born, the marsupials' birth process is merely another step in embryonic development. After between eight and forty-two days in the uterus, the marsupial fetus is "born," and then immediately deposited in the mother's pouch for several more weeks, until it is ready to survive outside of the maternal body.

A90

Bears are found on every continent except two. Which two?

Australia and Antarctica

A91

What is the name for the annual period of sexual interest among deer and elk?

The Rut

Creatures and Environments

Bird Migration

Most birds have eyes that are two or three times stronger than human eyes. They use their excellent sight and their innate power of orientation to migrate, often returning to the same neighborhood and sometimes even the same nest, year after year.

Some birds like pheasants and grouses migrate short distances, but most of our feathered friends fly thousands of miles during their seasonal migrations.

Q92 **What type of bird flies the farthest during its yearly migration?**

A. The sooty albatross
B. The arctic tern
C. The golden plover
D. The rufous hummingbird

Animal Sex

For some animals, sex is just a fleeting passion. Others take it much more seriously.

Q93 **Which of the following mammals forms stable nuclear families after mating?**

A. Jackals
B. Giraffes
C. Polar bears
D. Sea lions

A92

What type of bird flies the farthest during its yearly migration?

B is the correct answer. The arctic tern flies from the North Pole to the South Pole and back again with a total trip averaging around 24,000 miles. The greatest one-way distance covered by a ringed arctic tern was 14,000 miles flown in one voyage from north to south.

The sooty albatross, the largest of all ocean birds, continuously circles the world at 40 degrees south latitude and in doing so, has been known to cover a distance of 19,000 miles in 80 days, but strictly speaking continuous flight is not considered a migration.

The golden plover, another long-distance flyer, journeys no less than 12,000 miles in its yearly flights between the coast of Labrador and southern Brazil.

The tiny rufous hummingbird, weighing only 0.12 ounces, leaves Alaska every autumn for a trip along the coast to Mexico, 4,000 miles roundtrip, migrating farther than a large swan—a bird weighing 50 pounds, almost 7,000 times heavier than the tiny rufous.

A93

Which of the following mammals forms stable nuclear families after mating?

A is the correct answer. Jackals are one of the most maligned creatures—and one of the most committed to nuclear families. They often form life-long relationships with their mates—and jackal pups will stay with their families for two to three years.

Creatures and Environments

Animal Communication

Recent scientific studies indicate that man is not the only animal on earth to have developed a complex system of communication within his ecosystem.

Marine mammals communicate chiefly by sound, perhaps because of the poor visibility and the good sound-transmitting characteristics of water.

Q94 **Which of these undersea creatures change their complicated underwater communication patterns every few years?**

A. Sharks
B. Dolphins
C. Seals
D. Whales

Rivers

Rivers form by the accumulation of runoff water, groundwater and water from springs and small streams. Their importance to human life is inestimable.

Q95 **What is the longest river in the world?**

A. The Nile
B. The Amazon
C. The Congo
D. The Mississippi

A94

Which of these undersea creatures change their complicated underwater communication patterns every few years?

D is the correct answer. Whales, more specifically, humpback whales, completely change their elaborate, melodious underwater breeding songs every few years.

Shark communication patterns are not well understood but are very primitive by comparison.

Like whales, dolphins and seals also produce complicated underwater communication patterns, but those patterns do not change significantly over time.

A95

What is the longest river in the world?

A is the correct answer. The Nile is 4,145 miles long or about 100 miles longer than the Amazon.

The Amazon, however, is considered to be the world's "largest" river, discharging more than one-fifth of all fresh water that flows from the mouths of the world's rivers into the oceans.

The Congo is 2,718 miles long and the Mississippi is 2,340 miles long, flowing from Minnesota to the Gulf of Mexico.

The Great Barrier Reef

The Great Barrier Reef, a sea wilderness of coral reefs, cays, islands and underwater caverns, stretches 1,260 miles off the coast of eastern Australia.

Sharply dropping tides, monsoons and cyclones affect the reef's geography but never destroy it, for the Great Barrier Reef is a living, self-repairing entity. This reef, one of the world's largest organic structures, has been built chiefly by animals that are only one-half inch long.

Q.96 **What are these tiny animals called?**

A. Krill
B. Polyps
C. Crawling worms
D. Spider shells

Q.97 **Between what two countries does the Dead Sea lie?**

Q.98 **What is the name of the forest that endures natural cycles of growth and decay instead of human regulation?**

A96

What are these tiny animals called?

B is the correct answer. A polyp is a coral animal, soft and fleshy like the anemone. Found in all oceans of the world, polyps form an external cup-like skeleton of limestone that in most cases becomes attached to already existing coral rock. These polyps, loosely called coral, build continuously upward, averaging 0.5–2.8 centimeters per year, until the reef reaches the water's surface.

Crawling worms and spider shells are both sheltered by the coral framework as are millions of fish and an extraordinary diversity of marine life.

A97

Between what two countries does the Dead Sea lie?

Israel and Jordan

A98

What is the name of the forest that endures natural cycles of growth and decay instead of human regulation?

Climax Forest

Chimpanzees

It has been said that one result of Adam and Eve's banishment from the Garden of Eden was the loss of human communication with animals. Since then, in many different ways, humans have been trying to "talk to the animals." Washoe is one of the first chimpanzees to be taught a language for the purpose of communicating with humans.

Q 99 What language was she taught?

A. English
B. Morse code
C. Sign language
D. French

Animal Dependency

All animals take care of their babies after birth. But some live in harsh surroundings where their young must quickly be able to fend for themselves.

Q 100 Which of the following species becomes independent the fastest?

A. Wildebeest
B. Polar bear
C. Hippopotamus
D. Wolf

A99

What language was she taught?

C is the correct answer. In 1966, R. Allen and Beatrice Gardner taught Washoe, an infant then, Ameslan (American Sign Language). And since that time dozens of chimps and gorillas have been the subject of language experiments. What scientists are trying to determine is whether the chimps are actually communicating or just mimicking their trainers. New evidence supports the belief that the chimps are using the language to communicate. Roger Fouts, a former assistant of the Gardners and a psychologist at Central Washington University, has been caring for Washoe for nearly 10 years. In 1979, he placed a 10-month-old male chimp, called Loulis, with Washoe, a female, to see if the young chimp would learn sign language from his adopted mother. Within eight days, Loulis was beginning to imitate the signs Washoe was using. By the time he was five, Loulis had a vocabulary of 55 signs and was using them to communicate not only with Washoe but also with his playmates.

A100

Which of the following species becomes independent the fastest?

A is the correct answer. The wildebeest of southern Africa, which is the only animal of the four that is vulnerable to predators, can't always keep its young safe from the preying hyenas. Young wildebeests can outrun most predators by the time they are just two days old.

Tidal Waves

A tidal wave or "tsunami," meaning "overflowing wave" in Japanese, is a series of ocean waves of great height and length most often caused by an earthquake, volcano or an underwater mudslide. This natural phenomenon can be quite destructive: in 1883 after a violent volcanic explosion on the Indonesian island of Krakatoa, 36,000 people died and 300 towns on surrounding islands were destroyed by tidal waves. Nine hours later thousands of boats were lost along the coastlines of Australia and India.

Q 101 How are tidal waves forecast?

A. From ships
B. From airplanes
C. From seismological observatories
D. From lighthouses on the coast

Tidal Wave Speed

The record height of a tidal wave is 278 feet, which was reached in 1971 when a tsunami hit Ishagaki Island in the Ryukyu chain just south of Japan. Its force tossed an 850 *ton* block of coral more than 1.3 miles.

Q 102 When the crest of a tidal wave does hit land, at what speed is it usually travelling?

A. 10 miles per hour
B. 30 miles per hour
C. 120 miles per hour
D. 600 miles per hour

A101

How are tidal waves forecast?

C is the correct answer. Ships and airplanes cannot see tidal waves in the deep ocean because an actual wave (of which the highest ever recorded was 278 feet in 1971) does not form until the tsunami approaches land, at which point if you can see it from a lighthouse on the coast it is probably too late to escape it.

A102

When the crest of a tidal wave does hit land, at what speed is it usually travelling?

B is the correct answer. In the deep sea a tidal wave, which can reach lengths of 100 miles, may have a forward speed of 600 miles per hour, but as it nears the coastline the velocity of the wave diminishes as the wave height increases.

Creatures and Environments

Photosynthesis

Plants trap sunlight and make sugar in a process called photosynthesis.

This sugar supplies the energy for all life on earth. Organisms that don't carry out photosynthesis must get their energy from those that do.

Q103 Which of the following does not carry out photosynthesis?

A. Seaweed
B. Potato
C. Venus's-flytrap
D. Mushroom

Rain Forests

Within 100 years from now, at current deforestation rates, much of the world's primary tropical forests will have disappeared.

Rain forests, with their dense triple canopy of foliage, cover hundreds of thousands of acres in the hot and wet equatorial lands of Latin America, West Africa and Asia. Sixty to eighty inches of rain fall every year in these dense lands.

Q104 What are they most valued for?

A. Fertile soil
B. Diversity of animal and plant life
C. Natural aquifers
D. All of the above

A103

Which of the following does not carry out photosynthesis?

D is the correct answer. The mushroom is a fungus, not a plant, that gets its food from decaying plant matter. Although the Venus's-flytrap and similar plants eat insects, they make some of their food through photosynthesis. Both seaweed and potato plants make all their food by photosynthesis.

A104

What are they most valued for?

B is the correct answer. The layers of the tropical rain forest are home for all kinds of mammals, birds, trees, shrubs, ferns and mosses, many of which are found nowhere else in the world. Close to 300 species of hummingbirds, for example, are confined to the forests of South America.

The soil in rain forests is barely fertile. With so much rain most of its silicates and compounds are leached away.

Natural aquifers are found underground, and it is rainfall, not aquifers, that provide rain forests with their water.

Ice Ages

During the world's ice ages, great ice sheets moved across the world, lowering the earth's temperature. Plants and animals died and strange landforms began to appear. On the frozen tundra water seeped into the ground, froze, and created masses of ice that pushed the surface into domes.

Q 105 These domes still exist. What are they called?

A. Glaciers C. Pingos
B. Volcanos D. Black holes

Q 106 What is the name of the frozen layer that lies beneath the ground surface in frigid places like northern Alaska?

Q 107 What kind of ecosystem is the polar bear's natural habitat?

A105

These domes still exist. What are they called?

C is the correct answer. Conical hills called pingos are pushed up in permafrost by ice accumulating beneath the surface. Sometimes 150 feet tall and 1,800 feet in diameter, they may collapse into volcanic-like shapes when the ice melts. Pingos may be seen in arctic regions, and in Wales as well.

A glacier is a huge moving mass of ice originally made of compacted snow.

A volcano is a vent in the earth's crust through which lava and gases are ejected during eruption.

A black hole is the theoretical gravitational collapse of a star or collection of stars.

A106

What is the name of the frozen layer that lies beneath the ground surface in frigid places like northern Alaska?

Permafrost

A107

What kind of ecosystem is the polar bear's natural habitat?

Arctic Tundra

Creatures and Environments

Dinosaurs

For 150 million years, dinosaurs dominated the earth. Then suddenly they vanished.

Q 108 Which theory is currently thought to be the most likely explanation for the end of the Age of Dinosaurs?

A. Early man hunted them to extinction

B. They froze to death in an ice age

C. Their large, conspicuous eggs were eaten by mammals

D. Their food supply was destroyed when an asteroid hit the earth

Bio-Rhythms

Persistent rhythmicity is widespread in the animal and plant kingdom. In the animal world, environmental factors like light and temperature as well as perhaps the existence of an internal "bio-clock" affect daily skin color changes, wakefulness, hibernation and migration.

Most humans, like other animals, are awake during the day and generally abide by the daily 24-hour cycle.

Q 109 What rhythm describes this type of behavior?

A. Circadian

B. Diurnal

C. Endogenous

D. Aestivate hibernal

A108

Which theory is currently thought to be the most likely explanation for the end of the Age of Dinosaurs?

D is the correct answer. Sixty-five million years ago, during the Cretaceous period, is when many scientists believe the dinosaurs were wiped out. Either a volcanic eruption or an asteroid collision with the earth, raising a huge cloud of dust that blocked the sun for thousands of years, is what scientists think caused dinosaurs and other terrestrial-like animals to eventually die out. By the way, the natural phenomenon that made the dinosaurs extinct is thought to be the same as the so-called "nuclear winter" that might follow a thermonuclear war of even limited proportions.

A109

What rhythm describes this type of behavior?

A is the correct answer.

Diurnal is when animals are active at night.

Endogenous means originating and developing from within, without the influence of external factors. And aestivate hibernal is a fancy term for passing the hot summer months in a state of torpor.

Creatures and Environments

Platypus

When a British sea captain brought the skin of an Australian duck-billed platypus to the British Museum in 1798, the officials thought it was a hoax. Beaver-like fur, a long tail, webbed claws and a duck bill instead of a nose, all made for a strange looking animal.

Q110 Almost a century later zoologists learned something even stranger about this mammal. What was it?

The platypus . . .

A. Has no vertebrae

B. Has gills

C. Has feathers on its underbelly

D. Lays eggs

Q111 What is the animal kingdom's fastest runner?

Q112 What is the generic name for an animal that kills other animals for food?

A110

Almost a century later zoologists learned something even stranger about this mammal. What was it?

D is the correct answer. The platypus harkens back 150 million years, and aside from the spiny anteater, also from Australia, it is the only other mammal known to lay eggs. In the world of taxonomy, discovering this unique method of reproduction was important because it represented a separate line of evolution while indicating the development of reptiles into mammals.

A111

What is the animal kingdom's fastest runner?

Cheetah

A112

What is the generic name for an animal that kills other animals for food?

Predator

Antarctic Wildlife

Antarctica is the coldest, windiest, driest and most forbidding of all the seven continents. Even though the sun shines at the South Pole for as many hours as it does on the equator, the temperature never rises above freezing. Still it is home to a wide variety of animals, and even a few scientists.

Q113 What is the average annual snowfall that these animals and people must endure?

A. 48 inches   C. 2 inches
B. 100 inches   D. 14 inches

Nomenclature

King Peter Came Over From Germany Seeking Fortune.

Q114 This simple statement is a mnemonic device in which science?

A. Astronomy   C. Physics
B. Taxonomy   D. Geology

A113

What is the average annual snowfall that these animals and people must endure?

C is the correct answer. By definition, Antarctica is a desert. Because of its extreme cold, moisture is low, which explains the low snowfall. The wind whips the little snow around just like sand in the Sahara, and gives the impression of a blizzard. The snow rarely melts, and over the course of millions of years, it has become compacted into the present icecap.

A114

This simple statement is a mnemonic device in which science?

B is the correct answer. Taxonomy is the systematic arrangement of plant and animal organisms according to accepted diagnostic criteria that determine their assignment to one of the following classifications:

1. **K**ingdom
2. **P**hylum
3. **C**lass
4. **O**rder
5. **F**amily
6. **G**enus
7. **S**pecies
8. **F**orm

Changes in all ranks of classification do occur and new orders are frequently proposed.

Depending on a biologist's preference, bacteria, for example, may be classified as a class of fungi, a phylum of the plant kingdom, or a phylum of yet a third kingdom that includes all single-celled organisms.

Creatures and Environments

Animal Habitat

The musk-ox is neither ox nor musk, and its native name, "oomingmak," means "the bearded one."

At 700 pounds, the musk-ox is a big, fierce-looking animal. It has been domesticated, though only after a close call with extinction.

Q115 **Which area is the musk-ox's natural home?**

A. India

B. Tanzania

C. Arctic tundra

D. Australia

Q116 **What is the process by which animals with exterior skeletons shed skin prior to growth?**

Q117 **What plant substance reacts to a specifically colored light?**

A115

Which area is the musk-ox's natural home?

C is the correct answer. By 1954 the musk-ox had been completely eliminated from Alaska and was scarce in Canada and Greenland. Geographer John Teal, Jr. championed their cause and captured several of the wild animals to start a domesticated herd of musk-ox on his Vermont farm. Teal's aim was to begin commercial production of the fine, light underwool that protects the animal during arctic winters. After Teal's project was successful, the state of Alaska commenced a similar project, and the musk-ox has since been reintroduced there.

A116

What is the process by which animals with exterior skeletons shed skin prior to growth?

Molting

A117

What plant substance reacts to a specifically colored light?

Phytochrome

Boa Constrictor

Boa constrictors and their relatives are the only snakes that have claws left over from the lizards from which they evolved.

Approximately 20 inches long at birth, boas are born like mammals—without the protective covering of an egg.

An extra bone in the boa's jaw allows its mouth to open wide—wide enough to swallow animals considerably larger than its own head, including large lizards, birds, opossums, dogs and various rodents. Probably the biggest known animal ever eaten by a boa was the 30- or 40-pound ocelot found in the stomach of a 10-foot boa back in 1894.

There are cases of boas dying from ingested porcupine quills, but their powerful stomach juices generally enable them to digest just about anything—bones and all—within a few days. But they must lie very quietly while they do.

One of the amazing things about these snakes is how long they can go without eating anything—sometimes as much as one year.

Q118 How can boa constrictors live so long without food?

A. They can hibernate for many months when they can't find anything to eat

B. They sleep up to 20 hours a day

C. They are extremely efficient at getting all the energy from their food

D. They are cold-blooded and don't need much energy

A118

How can boa constrictors live so long without food?

D is the correct answer. Mammals use most of their energy keeping their body temperature at a steady, high level, which hot-blooded cells need to function. The boa doesn't do that; when it's cold, the boa just slows down a bit.

HISTORY AND PERSONALITY

Henry Ford

Henry Ford is a name that has long been synonymous with the automobile.

Early in this century, Ford's most significant innovation revolutionized automobile production.

Q119 What was Henry Ford's greatest automotive contribution?

A. The assembly line
B. The storage battery
C. The internal combustion engine
D. The multi-speed transmission

Periodic Table

Chemical elements consist of atoms of only one kind. By themselves or in combination, chemical elements constitute all matter.

One scientist devised the periodic table of elements when only 60 of 113 elements were known, leaving blank spaces on his chart where he expected the elements to be discovered to fall.

Q120 What scientist devised the chart now known as the periodic table of elements?

A. Murray Gell-Mann
B. Dimitri Mendeleev
C. Galileo Galilei
D. Max Planck

A119

What was Henry Ford's greatest automotive contribution?

A is the correct answer. The assembly line allowed the Ford Motor Company to mass-produce thousands of identical Model A automobiles. This development revolutionized not only the automobile industry, but industrial production in general, becoming the forerunner for today's fully-automated production lines.

A120

What scientist devised the chart now known as the periodic table of elements?

B is the correct answer. Dimitri Mendeleev, the Russian chemist, created the periodic table a century ago.

Murray Gell-Mann won a Nobel Prize for his work hunting quarks.

Galileo was an Italian astonomer of the 16th and 17th centuries.

German physicist Max Planck proposed in 1900 that energy exists in discrete units or quanta.

History and Personality

B. F. Skinner

B. F. Skinner is a world renowned psychologist. After graduate work at Harvard, he became one of the most influential and provocative researchers of the 20th century. One of his landmark studies involved the training of pigeons, and his book *Walden Two* attracted much attention.

Q121 But it was due to which of the following that Skinner gained the most notoriety?

A. His scandalous private life
B. His philanthropic contributions
C. His political aspirations
D. His design of baby furniture

Inventions

Pennsylvania, Maryland, New Jersey and New York owe many of their boundaries to one of the U.S.'s most celebrated surveyors.

Grandfather clocks, compasses, levels and transits bear his name. And he is credited with developing the first telescope in the U.S.

Q122 Who was this man?

A. Thomas Jefferson
B. Benjamin Franklin
C. David Rittenhouse
D. Patrick Henry

A121

But it was due to which of the following that Skinner gained the most notoriety?

D is the correct answer. Skinner first attracted national publicity with the special climate-controlled crib he made for his daughter in 1945. Although Skinner believes we are controlled by the environment, he is not a fatalist. He doesn't think we can change the way we act by just talking about our emotional conflicts. But he says we can reshape our behavior by changing the patterns of rewards in our environment.

A122

Who was this man?

C is the correct answer. David Rittenhouse was born outside Philadelphia in 1732 and died in 1796. He was much like Benjamin Franklin in his varied interests in technology, astronomy and natural philosophy. But, like Franklin, these interests did not supersede his politics, and he was also active during the American Revolution.

Thomas Jefferson was the third president of the U.S., noted for his interest in architecture and natural history.

Among other things, Benjamin Franklin is known for his many inventions, industriousness and homilies.

Patrick Henry is renowned for his patriotic fervor during the Revolutionary War and for his 1775 speech when he spoke the words, "Give me liberty or give me death".

Imhotep

An Egyptian scholar named Imhotep is credited with being the first known scientist.

Q123 Which of the following is Imhotep most famous for?

A. Designing the great "step pyramid"

B. Damming the Nile for irrigation purposes

C. Navigating by the stars

D. Hybridizing wheat

Q124 What sacred Inca city is most famous for its ruins?

Q125 During the late 18th century, what Englishman was the force behind the world's first cast-iron bridge?

A123

Which of the following is Imhotep most famous for?

A is the correct answer. Imhotep is credited with being a physicist, mathematician, engineer and architect. Imhotep's step pyramid of Zoser, Saqqara, was higher than any other building in 2620 B.C. and was the world's first large scale monument in stone. For his efforts and wisdom, Imhotep was deified and worshipped for centuries following his death.

Damming the Nile was most likely accomplished by Menes (3100 B.C.), the first king of unified Egypt. He originated the basin system of irrigation by building banks along the Nile to control flooding.

Herodotus wrote that the Phoenicians used the star now known as Polaris as a guide in navigation, but accuracy in celestial measurements at sea awaited the 18th century invention of the sextant.

Work in wheat hybridization is thought to have started in the 17th century, and it was Gregor Mendel who laid the foundation for the understanding of plant hybridization in the 18th century.

A124

What sacred Inca city is most famous for its ruins?

Machu Picchu

A125

During the late 18th century, what Englishman was the force behind the world's first cast-iron bridge?

Abraham Darby

History and Personality

Gradualism

The concept of "gradualism" has long been a cornerstone in the theory of evolution. It states that all species of plant and animal life, including humans, have descended gradually from ancestors with significantly different physical traits.

Q126 Who was responsible for researching and creating this theory?

A. J. B. de Lamarck

B. Charles Darwin

C. Niles Eldredge

D. Alick Isaacs

Antarctica

Antarctica is probably earth's least-understood continent.

Although explorers crossed the Antarctic Circle for the first time two centuries ago, it was not until 1911 that humankind first set foot on the South Pole.

Q127 Who was the first person to reach the South Pole?

A. Robert Falcon Scott

B. Richard Byrd

C. James Cook

D. Roald Amundsen

A126

Who was responsible for researching and creating this theory?

B is the correct answer. The 1859 publication *On the Origin of Species* was an important event in the history of science, as it changed the way people thought about themselves and their history. The core of Darwin's theory is natural selection—nature's process of selecting the strongest of a species to survive and thus insuring gradual evolutionary development.

In his theory of evolution, Lamarck postulated that environmental changes cause inheritable change in animals and plants.

Niles Eldredge is a paleontologist at New York's American Museum of Natural History who collaborated with Stephen J. Gould on the theory of punctuated equilibrium—a theory that explains how some species have evolved, not gradually, but in fits and starts.

British doctor, Alick Isaacs is credited with discovering interferon.

A127

Who was the first person to reach the South Pole?

D is the correct answer, Norwegian explorer Roald Amundsen, who reached the South Pole on December 14, 1911, 34 days before his British rival, Robert Falcon Scott. The Amundsen team survived their trek while the Scott team died of exposure just days after arriving.

Captain James Cook was the first explorer to cross the Antarctic Circle, while American Richard Byrd became the first man to reach the South Pole by aircraft on his 1929 solo flight.

Anthropology

Charles Darwin speculated that Africa might be the continent where man emerged. There is one man who, along with his family, spent over 50 years in eastern Africa proving Darwin's thesis. The family discovered the complete skull of the earliest known ape, as well the remains of *Homo habilus*, which at 1.75 million years old was—at the time of discovery—man's oldest known ancestor.

Q128 **Who was this famous physical anthropologist?**

A. Louis Leakey

B. Franz Boas

C. Derek Freeman

D. Vincent Sarich

A128

Who was this famous physical anthropologist?

A is the correct answer. Louis Seymour Bazett Leakey (1903–72) was the son of a British missionary in Kenya. After he received his Ph.D. from Cambridge, Leakey spent his entire professional life exploring East Africa for traces of *Homo sapiens* ancestry and ancient culture. No one has contributed more to the discovery of ancient man than Louis Leakey, his wife Mary and son Richard.

Vincent Sarich rocked the anthropological community with his molecular clock theory. Using biochemical methods, his analyses of ape and human blood led Sarich to conclude that humans diverged from apes only 5 million years ago, not 20 million which has been the generally accepted theory.

Born in 1858, Franz Boas, through his teachings and field work, had more influence than anyone in the U.S. on the development of anthropology as a field of study. While at the American Museum of Natural History, he provided funding for Margaret Mead's famous trip to Samoa.

Derek Freeman is an Australian anthropologist who has lived and worked in Samoa. He recently wrote a controversial book attacking Margaret Mead's methodology and conclusions in her anthropological study, *Coming of Age in Samoa*.

Albert Einstein

Albert Einstein had a dream. He wanted, very badly, to formulate a single ultimate theory that would explain the existence of all matter, all energy and all forces in the universe.

Q129 Which of Einstein's theories made that dream come true?

A. Principle of general relativity
B. Theory of supergravity
C. Theory of supersymmetry
D. None of the above

Q130 Amos Alonzo Stagg Field at the University of Chicago played a role for what scientific team?

Q131 Name the U.S. government agency formed in 1950 to support scientific research.

A129

Which of Einstein's theories made that dream come true?

D is the correct answer. Einstein failed to reach his goal. According to modern physicists, Einstein used in his equations only two forces of nature, electromagnetism and gravity, and ignored the "strong" and "weak" forces that explain the existence of quarks and the process of radioactive decay.

It was the principle of general relativity that introduced a new way of thinking into the world. From this mathematical formulation, Einstein proved that the velocity of light is not constant—that it depends on its proximity to a gravitational field.

The theories of *supergravity* and *supersymmetry* are theories that include all the forces of nature and are being worked on by scores of physicists all over the world. Huge machines, called accelerators, have been developed to create a force similar to the unified force physicists believe existed in the earliest moments of the universe.

A130

Amos Alonzo Stagg Field at the University of Chicago played a role for what scientific team?

The Manhattan Project

A131

Name the U.S. government agency formed in 1950 to support scientific research.

National Science Foundation

Gene Regulation

Forty years ago, this scientist proposed our current theory of gene regulation, but it is only in the last decade that her work has been acknowledged by academia.

In 1983, she won the Nobel Prize in physiology and medicine.

Q132 Who is this scientist?

A. Rosalyn Yalow
B. Helen Taussig
C. Barbara McClintock
D. Helen Caldicott

Louis Agassiz

Louis Agassiz was a 19th century Swiss-American naturalist. He spent the last 25 years of his life teaching a new generation of U.S. naturalists at Harvard University. He is renowned for being the most prominent of U.S. biologists to disagree with Darwin's ideas about evolution.

Q133 This reputation has unfortunately obscured his extensive and important work in what field?

A. Glaciation
B. Medicine
C. Botany
D. Speleology

A132

Who is this scientist?

C is the correct answer. Barbara McClintock presented the first evidence of "jumping genes," controlling elements that change their position on the chromosome. For this work, Dr. McClintock was first ignored, and even ridiculed, by her colleagues, but it ultimately won her the highest award for scientific achievement, the Nobel Prize.

The other three scientists are distinguished medical doctors: Dr. Yalow in nuclear medicine, Dr. Taussig in pediatric cardiology, and Dr. Caldicott in pediatrics and antinuclear activism.

A133

This reputation has unfortunately obscured his extensive and important work in what field?

A is the correct answer. At first, Agassiz did not believe that glaciers moved. To prove himself correct, he experimented on the glaciers of Switzerland and found that they *did*, in fact, move. Further study led him to later introduce to the world the concept of the Ice Age.

Photography

One man made the machines that show us the fluid beauty of a drop of milk and capture a speeding bullet in mid-flight. He helped us win a war by lighting up the French countryside and froze the first instant of an atomic blast.

His stroboscope rearranges bits of time and motion. His photography expands both the limits of our vision and knowledge of our world.

Q134 **Who is he?**
A. Edwin Land
B. George Eastman
C. Edward Muybridge
D. Harold Edgerton

Q135 In 1977, what human-powered aircraft won the Kremer Prize by flying a figure eight over flat terrain with a minimum height of 10 feet at the beginning and end of the course?

Q136 In what year did the Russian launch of the satellite *Sputnik* instigate the space race?

A134

Who is he?

D is the correct answer. Harold Edgerton, an engineer at the Massachusetts Institute of Technology (M.I.T.), showed us things we could never see before, even though most of them were happening right in front of our eyes.

Edwin Land developed a one-step process for developing and printing photos, the basis of Polaroid cameras.

George Eastman developed photographic film and made photography a hobby for the masses with his Kodak camera.

Edward Muybridge made the first projected movie—a horse galloping—in 1880. He used 24 different cameras to make the 24 frames of his movie.

A135

In 1977, what human-powered aircraft won the Kremer Prize by flying a figure eight over flat terrain with a minimum height of 10 feet at the beginning and end of the course?

The *Gossamer Condor*

A136

In what year did the Russian launch of the satellite *Sputnik* instigate the space race?

1957

Early Astronomy

This man was a great Greek scholar who lived in the 6th century B.C. During that era he was one of the few who studied the natural environment. He believed the heavenly bodies were brought into existence by the same process that formed the earth, and he accurately explained moon phases and eclipses. For knowledge such as this, he was brought to trial and accused of impiety and atheism. And this was just the beginning of a long history of science conflicting with religion.

Q137 Who was this man?

A. Anaxagoras   C. Pythagoras

B. Hippocrates   D. Aristotle

Q138 In 1798, what clergyman warned that unchecked human population growth would eventually outstrip the earth's available food supply?

A137

Who was this man?

A is the correct answer, Anaxagoras.

On the Greek island of Kos, Hippocrates founded one of the first schools of medicine. He taught his students to look upon disease as a purely physical phenomenon—not something that could be ascribed to the arrows of Apollo.

Pythagoras lived in the 4th century B.C. and is best known for working out the Pythagorean theorem that states that the square of the length of the hypotenuse of a right triangle equals the sum of the squares of the lengths of its sides.

Aristotle was one of the greatest thinkers of ancient times. His lectures are collected into 150 volumes. This represented a "one-man encyclopedia" of the knowledge of his times with writings on the subjects of natural philosophy, science, politics, literary criticism and ethics.

A138

In 1798, what clergyman warned that unchecked human population growth would eventually outstrip the earth's available food supply?

The Reverend Thomas Malthus

History and Personality

John James Audubon

John James Audubon is well-known in the bird world, and his name is synonymous with conservation.

Q 139 During his lifetime, what was Audubon's major contribution to ornithology?

A. Paintings of birds
B. His discovery of new bird species
C. His systematic banding of birds
D. Breeding rare birds in captivity

Scott's Folly

In 1910, British Navy Captain Robert Scott set out on a scientific expedition to Antarctica. But he soon discovered that his old Norwegian rival, Roald Amundsen, had already started a trek to the South Pole. Captain Scott's expedition soon became a race.

Amundsen had spent years training with Arctic Eskimos, while Scott's preparation for the trip proved tragically misguided.

Q 140 Which of the following reasons did *not* contribute to Scott's failed attempt to be first at the South Pole?

A. He brought tractors that broke down within a few miles of his base camp
B. He brought ponies that soon died in the Antarctic conditions
C. He brought an airplane that froze without ever leaving the ground
D. He brought an ill-trained crew that couldn't handle the sled dogs

A139

During his lifetime, what was Audubon's major contribution to ornithology?

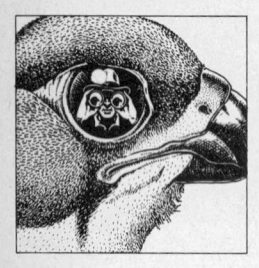

A is the correct answer. Audubon's paintings are indisputably beautiful, and the 1838 British publication of the 435 paintings making up *Birds of America* kindled wide interest in natural history. But for the most part, Audubon's interest in birds was artistic, not scientific.

The multivolumed *Dictionary of Scientific Biography* has this to say about a person we thought was one of the world's greatest ornithologists: "His artistic stature seems to dwarf his scientific stature, and the latter would probably still be less, had he not been a painter expected to provide text for his paintings. The chances seem very good that had he not been an artist he would be an unlikely candidate for a dictionary of scientific biography, if remembered to science at all."

A140

Which of the following reasons did *not* contribute to Scott's failed attempt to be first at the South Pole?

C is the correct answer. The Wright Brothers had not yet achieved flight when Scott reached the South Pole. All of the rest contributed to the deaths of Scott and his team when they tried to return to their ship after reaching the South Pole.

Watson and Crick

U.S. scientist James Watson and British scientist Francis Crick collaborated on an important biological discovery. They were awarded the Nobel Prize in chemistry (1962) and were praised by their colleagues.

Q141 What discovery did Watson and Crick make?

A. The structure of DNA
B. The birth control pill
C. The structure of protein
D. An artificial virus

A141

What discovery did Watson and Crick make?

A is the correct answer. With their colleague and sometime rival Maurice Wilkins, Watson and Crick discovered the structure of DNA, the molecule that carries genetic information and makes reproduction possible.

In the years since the structure of DNA was discovered, molecular biology has reached the point where it can do more than just identify certain structural subunits on the DNA molecule that correlate with certain morphological characteristics; it can make them happen through the technological process known as recombinant DNA. Scientists can now create entirely new forms of life by "splicing" existing life-forms together.

The birth control pill was developed by Gregory Pincus and John Rock.

Linus Pauling discovered the structure of protein.

And artificial viruses haven't been made by anybody—yet.

Nobel Prize

The first Nobel Prize was awarded in 1901. Alfred Nobel, a Swedish chemist and chemical engineer (he invented dynamite) bequeathed money for the start of the Nobel Foundation, which every year was to award prizes in the fields of physics, chemistry, medicine, physiology, literature, peace and later, economics.

Q142 Which one of the following persons has the distinction of being one of the few who have won two Nobel Prizes in the fields of science?

A. Kenneth Wilson
B. Linus Pauling
C. The International Red Cross
D. Marie Curie

Q143 The Geneva Protocol of 1925 banned the future deployment of what kind of warfare?

Q144 Who is known as the "father of the hydrogen bomb"?

A142

Which one of the following persons has the distinction of being one of the few who have won two Nobel Prizes in the fields of science?

D is the correct answer. Marie Curie shared the 1903 physics prize with her husband Pierre and Antoine Becquerel for the study of radiation phenomena and won the 1911 award outright for the discovery of two new radioactive elements. In later years her daughter, son-in-law and neighbor were also awarded Nobel Prizes.

Dr. Linus Pauling won two prizes, one in the field of chemistry and one in peace.

The International Committee of the Red Cross has won three peace prizes: in 1917, 1944 and 1963.

And Kenneth Wilson won the 1982 physics prize for his analysis of what basic changes occur in matter under the influence of temperature and pressure.

A143

The Geneva Protocol of 1925 banned the future deployment of what kind of warfare?

Chemical and Biological

A144

Who is known as the "father of the hydrogen bomb"?

Edward Teller

Quarks

The quest to search for the smallest unit from which all matter is made belongs to the physicist. First, an atom was discovered and then the nucleus, and finally the protons and neutrons that are just one ten-millionth of a centimeter in diameter.

Modern physicists now know that the proton can be broken down even further into a tiny particle called a quark.

Q145 Which of the following scientists has been a pioneer in the scientific understanding of these small atomic particles?

A. Noam Chomsky

B. Richard Feynman

C. Yuri Gagarin

D. Isaac Asimov

A145

Which of the following scientists has been a pioneer in the scientific understanding of these small atomic particles?

B is the correct answer. Richard Feynman is a professor of theoretical physics at the California Institute of Technology and has received a Nobel Prize for his work in quantum electrodynamics. One of Feynman's main interests has been developing, mathematically testing and understanding the consequences of the theory that explains quark behavior.

Of himself, Richard Feynman says, "I'm really still a very one-sided person and I don't know a great deal. I have a limited intelligence and I use it in a particular direction."

Noam Chomsky is a linguist whose revolutionary theories about the way language is learned radically altered the field of cognitive science.

Yuri Gagarin, a Russian, was the world's first human astronaut.

Isaac Asimov, born in 1920, is a U.S. biochemist and a prolific writer. He has written 250 fiction and nonfiction books on a plethora of subjects.

DNA Photographs

Watson and Crick won the Nobel Prize for their work in determining the "double helix" nature of the DNA molecule. Confirmation of their theory was aided by a series of X-ray diffraction photographs of the DNA molecule made under different conditions in humidity. These photos showed a helical pattern in the molecules.

Q 146 **Who was responsible for that series of remarkable photographs?**

A. Rosalind Franklin

B. Luis Alvarez

C. Walker Eastman

D. Andre Sakharov

A146

Who was responsible for that series of remarkable photographs?

A is the correct answer. Rosalind Franklin was a British industrial chemist and molecular biologist whose greatest strength lay in her technical innovations and employment of precise techniques on difficult macromolecules.

Luis Alvarez, a U.S. physicist, is most well-known for his work on radar, the atom bomb, the bubble chamber and for his 1980 finding of an unusually high concentration of the radioactive material iridium in a sedimentary core from southern Italy. The finding led to the theory that dinosaurs and the rest of life were annihilated in the wake of an asteroid falling to the earth and creating so much dust that all sunlight was blocked out.

Walker Eastman and the scientists in his laboratories simplified the photographic process and started the mass production of cameras and related equipment.

Andre Sakharov, Soviet physicist, worked on the hydrogen bomb, playing the role in the Soviet Union that Edward Teller had in the U.S. Starting in the 1960s, he spoke out against the use of nuclear weapons and won the 1975 Nobel Peace Prize. He is imprisoned in a provincial Soviet city.

Heart Transplants

In 1982, Seattle dentist Barney Clark became the first person ever to receive an artificial heart. Fifteen years earlier, in 1967, heart surgeon Dr. Christian Barnard performed the first successful heart transplant operation, in South Africa.

Q147 Who was the first U.S. surgeon to perform a successful heart transplant operation?

A. Dr. Denton Cooley
B. Dr. Robert Jarvik
C. Dr. Adrian Kantrowitz
D. Dr. Joseph Murray

Q148 What Frenchman is the father of craniofacial plastic surgery?

Q149 Whose arrival in 1778 marked the discovery of Hawaii for the Western world?

A147

Who was the first U.S. surgeon to perform a successful heart transplant operation?

C is the correct answer. Detroit's Dr. Adrian Kantrowitz performed the first U.S. heart transplant just days before Houston's Dr. Denton Cooley did. While Kantrowitz' patient lived for just seven hours after the operation, it was considered a "successful" procedure by much of the medical community at that time. Dr. Cooley's first patient lived for several weeks with a heart transplant, and contemporary heart transplant recipients enjoy a life expectancy of several years.

Dr. Robert Jarvik designed the artificial heart implanted in Barney Clark.

Boston surgeon Joseph Murray is famous for performing the world's first kidney transplant.

A148

What Frenchman is the father of craniofacial plastic surgery?

Paul Tessier

A149

Whose arrival in 1778 marked the discovery of Hawaii for the Western world?

Captain James Cook

Paleontology

Through his books, lectures and courses, including the hugely popular course titled "The History of Life on Earth" taught at Harvard, this scientist challenges people to rethink the popular biases in science, including the notions of progress, determinism, gradualism and adaptationism. Most recently he was invited by a group of students at the University of Witwatersrand in Johannesburg to lecture on human evolution and on the biological basis of human equality.

Q 150 Who is this famed paleontologist who claims as heroes his father, Joe DiMaggio and Charles Darwin?

A. Walter Alvarez
B. Stephen J. Gould
C. Alfred Russel Wallace
D. Jack Horner

A150

Who is this famed paleontologist who claims as heroes his father, Joe DiMaggio and Charles Darwin?

B is the correct answer. Early visits with his father to New York's American Museum of Natural History endowed Gould with a fascination with dinosaurs, while Joe DiMaggio taught him that "the thing that counts is excellence." Using the fossil records as valid evidence, Gould has stated that new species may arise abruptly and then settle down into a long period of stability. This theory, called "punctuated equilibrium" has challenged the scientific community and Darwin's idea, now entrenched, that most change in evolution is gradual.

Walter Alvarez is a geologist who along with a team of scientists has postulated that a large asteroid collided with the earth 65 million years ago, kicking up a thick layer of dust that blocked out the sun and caused many prehistoric animals—including dinosaurs—to become extinct.

Alfred Wallace, an Englishman, was a founder of modern evolutionary biology, along with Darwin. From his extensive field work in South America and Malaysia, Wallace made fundamental discoveries in biology, geology, geography and ethnography.

Jack Horner is an eminent paleontologist who unearths dinosaur "nests" in Montana. His fossil discoveries have led him to believe that dinosaurs may have been warm-blooded and not nearly as reptilian as generations of dinosaur scholars have thought.

Astronomy

This woman accomplished many "firsts." Among them, she was the first female member of the American Academy of Arts and Sciences and the first recognized female U.S. astronomer. The king of Denmark bestowed upon her a gold medal for the first discovery of a telescopic comet in 1847.

Q151 Who was she?

A. Maria Mitchell
B. Ida Tarbell
C. Ruth Benedict
D. Dorothy Hodgkin

Q152 In March 1978, what ship spilled 220,000 tons of crude oil into the Atlantic Ocean off the coast of Brittany, France?

A151

Who was she?

A is the correct answer. Maria Mitchell was born on the island of Nantucket where Quaker children "learned to box a sextant before they learned the queries of the Friends." There she "swept the skies" from a small rooftop observatory, daily logged her observations and calculations, and wrote articles for scientific journals. After several trips throughout the U.S. and abroad where her education and reputation grew, she was offered the position of professor of astronomy at the newly-formed Vassar College. She spent the last 25 years of her life there as a professor and as the director of the college's observatory, which is named after her.

Ida Tarbell was a journalist at the turn of the century who gained fame for a series of investigative articles on the great Standard Oil monopoly.

Ruth Benedict was an anthropologist who worked closely with Margaret Mead and wrote a landmark study of Japanese culture during World War II.

Dorothy Hodgkin won the 1964 Nobel Prize for her discovery of the structure of biochemical substances.

A152

In March 1978, what ship spilled 220,000 tons of crude oil into the Atlantic Ocean off the coast of Brittany, France?

Amoco Cadiz

History and Personality

Margaret Sanger

Margaret Sanger was born and brought up as a working class girl in upstate New York in the late 19th century. Before she died in 1966, Sanger had become internationally renowned.

Q153 For which of the following accomplishments did Margaret Sanger gain her fame?

A. The first Western woman to gain board certification in general surgery

B. The creator of the intelligence quotient or IQ test

C. The first female Nobel science laureate

D. The major force behind the development of the oral contraceptive

Sir Alexander Fleming

Sir Alexander Fleming made one of the major advances in 20th century medicine, but many are now saying that Fleming's innovation is being used *too* much.

Q154 What was Fleming's major contribution?

A. The smallpox vaccine

B. Penicillin

C. Anesthesia

D. Insulin

A153

For which of the following accomplishments did Margaret Sanger gain her fame?

D is the correct answer. Margaret Sanger was a leader in the movement toward birth control early in the 20th century, and her pioneering work stimulated the development of the oral contraceptive by some Massachusetts physicians in the 1950s.

A154

What was Fleming's major contribution?

B is the correct answer. Alexander Fleming discovered penicillin in 1928 when he noticed that bacteria couldn't grow near a green mold called penicillium. But some bacteria have developed a resistance to penicillin and overuse of the drug could magnify this problem.

The smallpox vaccine was developed by William Jenner in 1796. Two centuries later, in 1977, smallpox was virtually wiped from the face of the earth.

Anesthesia was introduced by U.S. dentists in 1846.

Insulin was isolated in 1921 by two Canadian doctors and immediately became essential medication for diabetes.

Seymour Papert

Seymour Papert, a mathematics and education professor at M.I.T., has said, "Education has very little to do with explanation. It has to do with engagement, with falling in love with the material."

With his colleagues at M.I.T.'s Artificial Intelligence Lab, Papert developed the only computer language specifically designed for learning. It is an easily accessible language that allows students of all ages to progress to the most sophisticated ideas in the world of programming.

Q155 What is the name of Papert's unique computer language?

A. Basic
B. Fortran
C. Logo
D. Lotus

Q156 In the 1950s, who revolutionized thinking about language acquisition by proposing that infants are born with an innate sense of language?

A155

What is the name of Papert's unique computer language?

C is the correct answer. The LOGO language does not require rote learning from its students. A graphic language, LOGO allows children to draw detailed, personal designs in a way that encourages the understanding of mathematics. With LOGO, there is no "right" or "wrong" so it is a language that provides an experience of exploration and discovery. Says Papert, "The spirit of LOGO is to produce a language that encourages an attitude of taking it and changing it, shaping it to yourself."

BASIC and FORTRAN are called "high level languages" that are used in the commercial and applied computation worlds. They are implementations of very early ideas from the 1950s about automatic programming being motivated more by arithmetic than by symbolic requirements.

LOTUS is the name of a software producing company in Cambridge, Massachusetts.

A156

In the 1950s, who revolutionized thinking about language acquisition by proposing that infants are born with an innate sense of language?

Noam Chomsky

Agrobiology

During the first half of the 20th century a Russian farmer propelled himself to fame by denouncing traditional genetic research and by starting a new field of "science" called agrobiology. He maintained that strains of wheat could be genetically altered by controlling the environment. This man developed theories he never proved, attacked Darwin's theory of natural selection and caused many highly reputable scientists to be imprisoned or executed.

Q157 Who was this man?

A. Nikola Tesla
B. Trofim Lysenko
C. Ivan Turgenev
D. Nikolay Vavilov

A157

Who was this man?

B is the correct answer. Although T. D. Lysenko held degrees from a Russian school of horticulture and the Institute of Agriculture, he effectively shut the door on agricultural research during his 25-year reign as director of the USSR's Institute of Genetics. He supported Soviet Darwinism, which held that the inheritance of characteristics could be directed by regulating the conditions of life, and it was this belief that became the official state science.

Nikola Tesla was a U.S. electrical engineer and inventor who made the use of alternating current practical (though only after a protracted battle with Thomas Edison who championed the less practical use of direct current).

Ivan Turgenev was a noted 19th century Russian novelist, poet, dramatist and critic.

Nikolay Vavilov, 1887–1943, was an excellent Russian botanist. He attempted to produce new strains of grain using Mendel's laws. In doing so, he ran afoul with Lysenko who forced his imprisonment, the conditions of which caused his death. In 1955 Vavilov was posthumously rehabilitated and honored.

Genetic Engineering

This man began his career as a physicist, but was attracted to biological research by James Watson, the co-discoverer of the double-helix structure for DNA. In 1955 the two met in Cambridge, England.

He held the American Cancer Society professorship at Harvard University for 25 years until his resignation in 1982 and won a share of the 1980 Nobel Prize in chemistry for his contributions to working out a research technique for the rapid sequencing of the subunits of DNA.

Q158 **Who is this scientist?**
A. Baruch Blumberg
B. Stephen Weisskopf
C. Walter Gilbert
D. Philip Morrison

Q159 **The nomadic Waorani tribe is isolated in the tropical rain forest of what South American country?**

Q160 **"Punctuated equilibrium" is an expansion of whose original evolutionary ideas?**

A158

Who is this scientist?

C is the correct answer. Walter Gilbert, who recently resigned from the genetic engineering company he co-founded at the beginning of the recombinant DNA revolution, remains a pioneer in the business of finding practical applications for genetic engineering.

Baruch Blumberg won the Nobel Prize in 1977 for his discovery of the hepatitis B virus.

Stephen Weisskopf is a molecular biologist and Dr. Gilbert's former business partner.

And Philip Morrison is a professor of astrophysics at M.I.T.

A159

The nomadic Waorani tribe is isolated in the tropical rain forest of what South American country?

Ecuador

A160

"Punctuated equilibrium" is an expansion of whose original evolutionary ideas?

Charles Darwin

Psychiatry

The 20th century has brought many changes to the world of psychiatry. Diagnoses and treatments have been developed where once none existed, and whole new fields of study have emerged; one of these is the care of dying patients.

Q 161 **Which internationally acclaimed psychiatrist has worked for more than 25 years fighting "the last and greatest taboo of the Western world, our fears of death and dying"?**

A. Elizabeth Kübler-Ross

B. Erik Erikson

C. Bruno Bettleheim

D. Anna Freud

A161

Which internationally acclaimed psychiatrist has worked for more than 25 years fighting "the last and greatest taboo of the Western world, our fears of death and dying"?

A is the correct answer. Swiss born, Elizabeth Kübler-Ross trained in Europe and joined medical relief work during World War II. After becoming a physician in 1957, she joined her American husband in the U.S. where she has been working ever since. Her books, lectures and indefatigable work with dying people has begun to change not only the health care profession, but the basic values of society as well.

Erik Erikson, now in his eighties, has concentrated on all aspects of human development. A renowned psychoanalyst, professor and author, Erikson is most well-known for his concept of the "eight stages of man."

Bruno Bettelheim, a contemporary of Erikson's, is also an acclaimed psychologist. Included in his list of books are: *Love is Not Enough—Treatment of Emotionally Disturbed Children*, *The Children of the Dream*, and *The Uses of Enchantment*.

Anna Freud, daughter of psychiatrist Sigmund, is considered a leading authority of orthodox analytic views. She was a major contributor and guide in the development of child analysis.

History and Personality

Gail Borden

Tang, Jello and frozen orange juice all owe their development to Gail Borden, the grandaddy of the instant food movement. For miners heading West during the 1849 gold rush, he manufactured a terrible-tasting dried beef biscuit that would not spoil, and he was the first to commercially extract juice concentrates. His most famous product borrowed technology from the Shakers and achieved success during the Civil War.

Q162 What was it?

A. Powdered milk
B. Instant coffee
C. Instant oatmeal
D. Condensed milk

Thomas Edison

Thomas Alva Edison is probably the most famous inventor in world history.

Q163 Which of the following machines is *not* an Edison invention?

A. The phonograph
B. The telegraph
C. The fluoroscope
D. The ticker tape

A162

What was it?

D is the correct answer. The Shakers used a vacuum pan to preserve fruits. Gail Borden travelled east to borrow their equipment and set up shop condensing milk. Although he thought his milk remained fresh because it was condensed, it is now known that the temperature he used destroyed the disease-carrying bacteria in milk. The Union Army made condensed milk a staple in the soldiers' diet, insuring Borden of financial success, which later allowed the company to expand and to eventually produce instant coffee and powdered milk.

A163

Which of the following machines is *not* an Edison invention?

B is the correct answer. The telegraph was invented before Edison's birth. Edison, however, did master the telegraph as a teenager when he gained employment working as a telegrapher on a midwestern railroad. That ultimately led him to invent the "ticker tape," a version of the telegraph that printed price quotations from the New York Stock Exchange onto a moving paper tape.

PHYSICAL SCIENCES

Physical Sciences

Mount St. Helens Energy

Volcanic eruptions unleash enormous geothermal power from deep within the earth.

Mountains have been destroyed and new islands created by volcanos in some of the most spectacular and violent displays in nature.

Q164 If all of the power of the first major Mount St. Helens eruption in 1980 could have been converted into electricity, it would have satisfied U.S. electric power demands for how long?

A. 5 weeks
B. 12 days
C. 12 hours
D. 3 months

Q165 What is the porous, glassy rock formed by volcanic eruptions?

Q166 What gases combine with water in the atmosphere to create acid rain?

A164

If all of the power of the first major Mount St. Helens eruption in 1980 could have been converted into electricity, it would have satisfied U.S. electric power demands for how long?

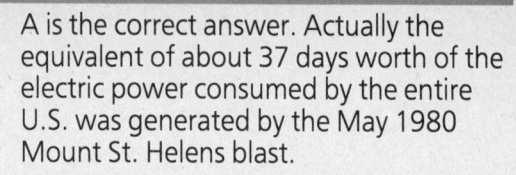

A is the correct answer. Actually the equivalent of about 37 days worth of the electric power consumed by the entire U.S. was generated by the May 1980 Mount St. Helens blast.

When Mount St. Helens erupted in May, for example, two-thirds of a cubic mile of volcanic rock was displaced, a blanket of volcanic ash was spread over four states, one billion dollars' worth of property was destroyed, and 62 people were reported killed or missing.

Put another way, the sustained power output exhibited from Mount St. Helens in May 1980 was about the equivalent of detonating 27,000 Hiroshima bombs—one per second over a seven and one-half hour period.

A165

What is the porous, glassy rock formed by volcanic eruptions?

Pumice

A166

What gases combine with water in the atmosphere to create acid rain?

Sulfur Dioxide and Nitrous Oxide

Statistics

Statistics is the science that deals with the collection, analysis, interpretation and presentation of numerical data like the following set of numbers:

1, 1, 1, 2, 3, 3, 3

Q167
Which one of the following quantifiers does this set of numbers lack?

A. Mean
B. Median
C. Mode
D. Range

Mathematics

There are several classical unsolved problems of pure mathematics. And they have resisted solution by the world's greatest mathematical minds.

In 1976, one of these problems was solved. The problem required a proof for the lowest number of colors that would be sufficient to color any map on a plane or sphere so that no two adjacent areas have the same color. This must hold true not just for the maps we see in an atlas, but for any conceivable map.

Q168
What are the fewest number of colors needed to color any map so that no two adjacent areas have the same color?

A. 3 colors
B. 4 colors
C. 6 colors
D. 8 colors

A167

Which one of the following quantifiers does this set of numbers lack?

C is the correct answer. The mode is defined as the most frequent number in a statistical distribution. In this case, both the numbers 1 and 3 appear the same number of times (3). Since neither appears more frequently, there is no mode.

The mean is the average of these numbers (14 divided by 7, or 2), the median is the middle number in the series (2), and the range is the amount between the first and last numbers (3 minus 1, or 2).

A168

What are the fewest number of colors needed to color any map so that no two adjacent areas have the same color?

B is the correct answer. The problem has been called the four-color problem. It was solved with the aid of a computer, which participated by drawing thousands of configurations. But many mathematicians would argue that what the computer did was not a true mathematical proof.

Physical Sciences

Greenhouse Effect

The greenhouse effect describes what some scientists believe to be a build-up of CO_2 in the earth's atmosphere. Now, in addition to natural CO_2 fluctuations, humans are responsible for adding on a yearly basis 18 billion tons of CO_2 to the earth's atmosphere. The CO_2 molecule does not absorb the heat radiated from the sun to the earth, but it is capable of absorbing heat radiated from the earth back into the atmosphere, thus the potential for dangerous worldwide climatic changes.

Q169 Which of the following does not noticeably change the amount of CO_2 in the earth's atmosphere?

A. Cement production
B. Desert expansion
C. Fossil fuel burning
D. Asbestos mining

Q170 What term describes a system's relative alkalinity or acidity?

Q171 What element constitutes about three-fourths of the earth's atmosphere?

A169

Which of the following does not noticeably change the amount of CO₂ in the earth's atmosphere?

D is the correct answer.

CO_2 is removed from limestone in the production of cement, contributing two percent of anthropogenic CO_2 in the atmosphere.

Large scale changes in the earth's terrestrial vegetation also cause fluctuations; so with fewer plants, there would be less CO_2.

Fossil fuel burning adds most of the 18 billion tons of CO_2 per year, though only about half of that remains in the air. The rest, scientists believe, is absorbed by the oceans.

A170

What term describes a system's relative alkalinity or acidity?

pH value

A171

What element constitutes about three-fourths of the earth's atmosphere?

Nitrogen

Atoms

Every element is made up of molecules that are, in turn, made up of atoms.

Atoms are described by their electrical charge as well as by their weight.

Q172 **Which parts of an atom determine an element's "atomic number"?**

A. Electrons
B. Protons
C. Neutrons
D. Ions

X-ray Radiation

Radioactivity is the spontaneous disintegration of an atom's nucleus accompanied by the emission of radiation.

X-rays are one form of emission by radioactive substances.

Q173 **Which type of radiation is an X-ray?**

A. Alpha
B. Beta
C. Gamma
D. Delta

A172

Which parts of an atom determine an element's "atomic number"?

B is the correct answer. The tiny nucleus of an atom contains all the protons and neutrons of that atom; on the outskirts of the atom lie the electrons. The "atomic number" of an element represents the number of protons found in the nucleus of an atom, while the "atomic weight" of an element is the number of protons and neutrons combined. Since electrons have a negligible mass, their number is not included in the calculation of "atomic weight."

A173

Which type of radiation is an X-ray?

C is the correct answer. X-rays are the radiation of gamma rays and are much more penetrating than either alpha (or anode) rays, which are composed of positively charged particles; beta (or cathode) rays, which are composed of negatively charged electrons; or delta rays, which are electrons ejected by an ionizing particle in its passage through matter.

Wilhelm Konrad Roentgen won the Nobel Prize in physics when he discovered the X-ray in the early 20th century.

Rock Types

In all the world's geological formations, rocks can all be separated into three categories (depending on how the rocks originated): sedimentary, igneous, metamorphic.

Q174
Which of the following is an example of igneous rock?

A. Granite

B. Coal

C. Limestone

D. Marble

Q175
What scientific term describes the pressure- and stress-related generation of electric impulses within crystalline rock?

Which of the following is an example of igneous rock?

A174

A is the correct answer. Igneous rocks crystallize from molten materials, either within the earth or at the earth's surface. Granite is an example of intrusive igneous rock, which is characterized by coarse-grained textures caused by slow cooling during their formation.

Sedimentary rocks, of which coal and limestone are examples, originated from compacted material that had been separated from preexisting rocks and transported to its final resting place by water, wind or ice. Because sedimentary rocks form at normal temperatures and pressures at the earth's surface, they are generally the only rocks that contain fossils.

Marble is an example of metamorphic rock, which is a former sedimentary or igneous rock that has been recrystallized, usually by heat, mineralized solutions and the pressure of deep burial. Other examples are slate and soapstone.

What scientific term describes the pressure- and stress-related generation of electric impulses within crystalline rock?

A175

Piezoelectric Effect

Physical Sciences

Blue Sky

Children want to know everything about the world around them. Often their simplest questions are the hardest to answer.

Probably the classic child's question is, "Why is the sky blue?"

Q176 We know that light of all colors pours out of the sun, but why do we only see blue sky?

  A. The nerve cells in our eyes are most sensitive to the blue light
  B. Hot plasma on the sun's surface scatters blue light
  C. The earth's magnetic field bends blue light more than yellow or red
  D. Tiny clumps of air molecules deflect blue light

Q177 What is the name of the turbulent air that trails an airplane during takeoff?

Q178 What space is entirely devoid of matter?

A176

We know that light of all colors pours out of the sun, but why do we only see blue sky?

D is the correct answer. The sky is blue because tiny clumps of air molecules bouncing randomly through the atmosphere deflect and scatter waves of blue light.

Waves of red, yellow and green light are much too long to be affected by the tiny air molecules, so they pass right through.

But waves of blue light are short enough to be scattered by the air molecules, making the entire sky seem blue on a cloudless day.

A177

What is the name of the turbulent air that trails an airplane during takeoff?

Vortex

A178

What space is entirely devoid of matter?

A vacuum

Galaxy

A galaxy is a system of stars, nebulae and interstellar gas and dust. Our solar system is but a very small part of the great spiral galaxy, the Milky Way. The advent of radio astronomy, a technique by which electromagnetic radiation, rather than visible wavelengths, is measured, has enabled astronomers to approximate the number of stars in our galaxy.

Q179 Approximately how many stars are in the Milky Way?

A. 50 million
B. 100 million
C. 50 billion
D. 100 billion

Halley's Comet

Halley's Comet was seen and recorded as far back as 240 B.C., but in 1684 its path was charted, and so named, by the Englishman Edmund Halley who demonstrated that comets travel in orbits around the sun.

During the first two weeks of April 1986, Halley's comet will be visible to the naked eye in areas of the southern U.S.

Q180 When gazing upon the comet, exactly what will people be looking at? A comet made up of . . .

A. Rocks and gases
B. Ice, water and rock particles
C. A combination of gases
D. None of the above

A179

Approximately how many stars are in the Milky Way?

D is the correct answer. The Milky Way contains more than 100 billion stars. The stars are arranged in a flattened spiral that slowly revolves—one turn taking about 200,000 years.

A180

When gazing upon the comet, exactly what will people be looking at? A comet made up of...

B is the correct answer. Astronomer Fred Whipple calls this comet a "dirty snowball" since its nucleus is made up of water, ice and tiny bits of rock.

Physical Sciences

Codes

Cryptography can be called the science of secrecy because it deals with the creation of codes.

All but one of the following numbers share a particular characteristic of divisibility that makes them useful to cryptographers.

Q181 **Which number does not share this characteristic?**

A. 2　　　　C. 13
B. 5　　　　D. 21

Q182 **What do geologists call the ancient continent that once included Africa, South America, Australia, Antarctica and India?**

A181

Which number does not share this characteristic?

D is the correct answer. The number, 21, is not useful because it can be divided by other numbers besides itself and one. The other numbers—2, 5 and 13—share the characteristic of being divisible only by themselves and by one. These numbers are called primes.

There are an infinite number of prime numbers. Prime numbers are useful to codemakers because even the highest-powered computers cannot break codes that are composed of two or more very long prime numbers multiplied together.

A182

What do geologists call the ancient continent that once included Africa, South America, Australia, Antarctica and India?

Gondwanaland

Weather

A strong low pressure system about 300 miles west of a southern city is getting stronger. The barometer is falling fast to below 29 inches—that's almost a record. And an unusually fast jet stream with winds of 150 miles per hour exists. Drastic differences in temperatures surround a cold front covered by a dense clouds that is coming in with the low pressure system.

Q183
These developments were a portent of what type of weather the following day, March 28, 1984, in Raleigh, North Carolina?

A. Fair, but with extremely strong winds
B. A hurricane
C. A tornado
D. A spring blizzard

Absolute Zero

Heat can be defined as the energy associated with the random motions of molecules, atoms or smaller structural particles of which matter is composed.

In theory, there is a temperature at which there is no molecular motion and therefore no heat. This temperature is called absolute zero.

Q184
On the Fahrenheit scale, what temperature is absolute zero?

A. −156 degrees
B. −273 degrees
C. −459 degrees
D. −707 degrees

A183

These developments were a portent of what type of weather the following day, March 28, 1984, in Raleigh, North Carolina?

C is the correct answer. A tornado. Hurricanes form offshore, and it was too warm for snow, although it did snow 300 miles to the northwest of Raleigh. But the weather conditions the night before were perfect for very severe thunderstorms, and tornadoes are always a possibility then.

A184

On the Fahrenheit scale, what temperature is absolute zero?

C is the correct answer. Minus 459 degrees F. Absolute zero is strictly a theoretical limit and cannot be achieved by any technological process.

Minus 156 F is the lowest temperature ever recorded on earth.

Absolute zero is minus 273 Celsius.

Minus 707 F is theoretically impossible to achieve.

Earthquakes

Hundreds of earthquakes occur throughout the world each year. While most earthquakes are barely perceived by humanity, others cause huge disasters.

Beside its relative seismological magnitude, the damage caused by an earthquake is determined by the depth at which the quake starts, its location and the quality of local building materials.

Q185 **Of the following earthquakes, which was rated highest on the Richter Scale?**

A. The Alaskan earthquake of 1964, 131 people killed

B. The Nicaraguan earthquake of 1931, 2,450 people killed

C. The Chinese earthquake of 1976, 655,235 people killed

D. The Calcutta earthquake of 1737, 300,000 people killed

Q186 **What kind of star suddenly becomes thousands of times brighter and then fades?**

Q187 **What region in the Atlantic Ocean has inspired some to suggest that paranormal forces are causing ordinary objects to vanish?**

A185

Of the following earthquakes, which was rated highest on the Richter Scale?

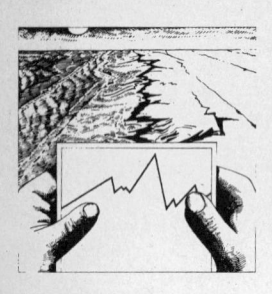

A is the correct answer. The Alaskan earthquake of 1964 occurred in a relatively unpopulated area, but it was rated 8.94 on the Richter Scale.

The Chinese quake, also very strong, was rated 8.2.

The Nicaraguan quake was "only" a 5.6, but because that quake was close to the surface, and because the buildings were not sturdy, the strength of the quake's energy inflicted great damage.

The Richter rating of the Calcutta quake of 1737 is, of course, unknown because the Richter Scale was not developed until the early 20th century.

A186

What kind of star suddenly becomes thousands of times brighter and then fades?

Nova!

A187

What region in the Atlantic Ocean has inspired some to suggest that paranormal forces are causing ordinary objects to vanish?

Bermuda Triangle

Mathematical Principle

Q188 The fact that an object displaces its density times its volume times its gravitational force is . . .

A. The Peter principle
B. Bernoulli's principle
C. Pascal's principle
D. Archimedes' principle

Q189 What is the name for the random movement among microscopic particles suspended in fluid?

A188

The fact that an object displaces its density times its volume times its gravitational force is . . .

D is the correct answer. Third century mathematician Archimedes discovered the principle when called upon by King Hiero of Syracuse to determine whether a certain gold crown was adulterated by silver.

The Peter Principle put forth by educator Laurence Peter says that in a hierarchy, every employee tends to rise to his level of incompetence.

In the 18th century Daniel Bernoulli made a statement about conservation of energy for fluids in a steady flow—that the sum of the energy of velocity, the energy of pressure and the potential energy of elevation remains constant.

And Pascal's principle (one of his many mathematical discoveries) states that pressure exerted on a fluid in a closed vessel is transmitted undiminished throughout the fluid. The hydraulic press is based on this principle.

A189

What is the name for the random movement among microscopic particles suspended in fluid?

Brownian Movement

Numbers

In 300 B.C. the Greek scholar Eratosthenes proved that there exists an infinite amount of numbers divisible only by themselves and one: 1, 2, 3, 5, 7, 11, 13 and so on.

Q 190 What do modern mathematicians call these numbers?

A. Square numbers

B. Prime numbers

C. Rational numbers

D. Composite numbers

Q 191 What is Avogadro's Number?

A190

What do modern mathematicians call these numbers?

B is the correct answer. Eratosthenes followed in Euclid's footsteps at the school of Alexandria—the greatest mathematical school of ancient times. He used a "sieve" as a method of sifting out the composite numbers in the natural series, leaving only prime numbers behind. Today, this system is still called the "sieve of Eratosthenes" and remains in use, for no general formula for the derivation of primes has been discovered.

Square numbers are the products of numbers multiplied by themselves.

Rational numbers are integers and their quotients.

Composite numbers are those numbers composed of one or more prime number factors.

A191

What is Avogadro's Number?

6.02 times 10^{23} molecules per mole

Physical Sciences

Planets at Midnight

Because of their complex movements through the field of stars in the night, the planets have fascinated people since long before anyone knew what they were.

We know of nine planets: Mercury, Venus, Earth, Mars, Jupiter, Saturn, Uranus, Neptune and Pluto.

The positions of the eight planets in our sky depends on where they are in their orbits compared with where our earth is.

Q192 Which of the following planets is never visible from the earth at midnight?

A. Venus   C. Jupiter
B. Mars   D. Saturn

Q193 What kind of land area drains into a single water course or body of water?

"My Very Earnest Mother Just Served Us Nine Pickles" is a traditional mnemonic device in astronomy. What does it help to remember?

A192

Which of the following planets is never visible from the earth at midnight?

A is the correct answer. At midnight, we can only look along the earth's orbit, and farther out. We can't look toward the sun, because the earth is in the way. Venus's entire orbit is closer to the sun than ours. It is often visible before the sun rises or after the sun sets, but never at midnight.

The same is true of Mercury.

A193

What kind of land area drains into a single water course or body of water?

Watershed

A194

"My Very Earnest Mother Just Served Us Nine Pickles" is a traditional mnemonic device in astronomy. What does it help to remember?

The order of planets away from the sun, starting with Mercury

Physical Sciences

Fahrenheit/Celsius

Temperature is measured on one of several arbitrary scales such as the expansion of mercury.

The two most common temperature standards are the Fahrenheit and Celsius scales.

Q195 At what point does the temperature Fahrenheit equal the temperature Celsius?

A. +32 degrees
B. −16 degrees
C. −100 degrees
D. −40 degrees

Atmospheric Weather

The earth's atmosphere is made up of a variety of gases, vapors and suspended materials that are bound to the earth by gravitational force. The diurnal and seasonal changes in the atmosphere dramatically influence the daily weather. But it is the position of the sun that is the main cause of the many fluctuations.

Q196 If you were an airplane pilot, when would you consistently expect the wind to adversely affect your flying?

A. Early morning
B. Midday
C. Late afternoon
D. Early to late evening

A195

At what point does the temperature Fahrenheit equal the temperature Celsius?

D is the correct answer. Minus 40 degrees. The conversion from Celsius to Fahrenheit can be accomplished with the following formula: Temperature Celsius equals 5/9 Temperature Fahrenheit minus 32. The number that makes TC equal to TF in this equation is −40.

A196

If you were an airplane pilot, when would you consistently expect the wind to adversely affect your flying?

C is the correct answer. The wind generally increases during the day and dies down by nighttime.

Organic Chemistry

Organic chemistry is the branch of chemistry that deals chiefly with hydrocarbons and their derivatives.

The simplest organic compound is a carbon atom surrounded by four hydrogen atoms. Its molecular formula is CH_4.

Q197
What is the name of the CH_4 molecule?

A. Ethane

B. Methane

C. Propane

D. Butane

Q198
In wine-making, what is the name for the chemical process by which sugars are converted to alcohol, CO_2 and, ultimately, vinegar?

Q199
What is the superheated matter of a fusion reaction called?

A197

What is the name of the CH₄ molecule?

B is the correct answer. Methane, a colorless, odorless, flammable gas, is lighter than air and forms explosive mixtures with air or oxygen. Methane occurs naturally as a product of the decomposition of organic matter in marshes, mines and especially in natural gas.

Ethane, naturally occurring but also produced as a by-product of petroleum refining, is C_2H_6.

Propane C_3H_8 also occurs naturally and is used chiefly as a fuel and as an ingredient in manufacturing chemicals.

Butane is C_4H_{10}, a common industrial fuel not naturally occurring but obtained by cracking petroleum and liquifying natural gas.

A198

In wine-making, what is the name for the chemical process by which sugars are converted to alcohol, CO_2 and, ultimately, vinegar?

Oxidation

A199

What is the superheated matter of a fusion reaction called?

Plasma

Physical Sciences

Geometry

Geometric shapes with four sides and four corners are called quadrilaterals.

One kind of quadrilateral is called a trapezoid.

Q200 Which one of the following shapes is a trapezoid?

A.
B.
C.
D.

Q201 In mathematics, what Greek letter signifies the ratio of the circumference of a circle to its diameter?

Q202 What measurement specifies the degree of disorder or randomness in a system?

A200

Which one of the following shapes is a trapezoid?

A. ▱

B. ▱

C. ▱

D. ▱

A is the correct answer. Trapezoids are quadrilaterals with only two parallel sides.

B is not a trapezoid because none of its sides are parallel; B is called a trapezium.

C is not a trapezoid because both pairs of its opposite sides are parallel. C is called a parallelogram.

D is not a trapezoid because both pairs of its opposite sides are parallel and all of its sides are equal in length. D is called a rhombus.

A201

In mathematics, what Greek letter signifies the ratio of the circumference of a circle to its diameter?

Pi

A202

What measurement specifies the degree of disorder or randomness in a system?

Entropy

Physical Sciences

Supernova

"It brightens until it is as luminous as a billion suns."

Such was the statement in 1054 when a famed supernova was observed by Oriental astronomers. It was visible to the naked eye for two years, and its remnants today are known as the Crab Nebula.

Q 203 **What is a supernova?**

A. A galaxy
B. A universe
C. A new star
D. An exploding star

Q 204 **Halley's Comet became visible in late 1985 for the first time in how many years?**

A203

What is a supernova?

D is the correct answer. "Supernova" is actually a misnomer since it literally means "new star"; a supernova is really the explosion (a gravitational collapse) of an *old* star in its terminal stage of its evolution. It may be, at maximum, tens of millions of times brighter than the original star and may even outshine the entire galaxy in which it explodes. The brightness of most supernovas peaks in a few days and then declines over a period of a few months.

It is easier to see supernovas in nearby galaxies, for in our own they are obscured by dust. They are thought to happen only once in every century.

A galaxy is a system of stars, nebulae and interstellar gas and dust.

The term universe applies to all matters of existence from subatomic particles through the largest aggregates known to man. The study of our universe implies the study of our galaxy, the Milky Way, and other galaxies.

A204

Halley's Comet became visible in late 1985 for the first time in how many years?

76 years

Cosmology

A raisin cake is rising in the oven. As the cake expands, each raisin becomes farther away from all the other raisins. The rate at which the farthest raisins recede is proportional to their distance from an initial raisin.

Q.205 This analogy is often used to explain a law that states that the universe is expanding uniformly. What law is that?

A. Hubble's constant
B. Boyle's law
C. The law of large numbers
D. Ohm's law

Q.206 In the Einstein equation $e = mc^2$, what does "c" represent?

A205

This analogy is often used to explain a law that states that the universe is expanding uniformly. What law is that?

A is the correct answer. Hubble's law states that the distance to a galaxy is directly correlated with the speed of its recession. By following this law backwards, astronomers came up with the "Big Bang" theory that postulates that all matter in the universe was together at some point in time before a huge explosion occurred.

Boyle's law describes the empirical relationship between the volume and pressure of a gas at a constant temperature.

The law of large numbers is a basic law of probability and statistics. It relates the probability of the event to the frequency of the occurrence of the event.

Ohm's law is an empirical rule relating to electrical current and is useful in the design and analyses of electrical circuits.

A206

In the Einstein equation $e=mc^2$, what does "c" represent?

The speed of light

Physical Sciences

Sound

The rigidity of a material governs the ability of sound to travel through it.

Q207 In which of the following materials does sound travel the fastest?

A. Steel
B. Quartz
C. Gold
D. Sea water

Q208 Name the science of sound.

Q209 How long does sunlight take to reach the earth?

A207

In which of the following materials does sound travel the fastest?

B is the correct answer. Sound travels 18,000 feet per second in quartz. In steel, sound travels 16,000 feet per second; in gold, 5,717 feet per second; and in sea water, only 4,800 feet per second.

Sound travels faster in liquids than in gases and faster still in solids.

A208

Name the science of sound.

Acoustics

A209

How long does sunlight take to reach the earth?

Eight minutes

Physical Sciences

Planets

Nine planets are in our solar system.

In order of their distance from the sun, they are Mercury, Venus, Earth, Mars, Jupiter, Saturn, Uranus, Neptune and Pluto.

Q210 Which planet has the highest daily temperature?

A. Mercury
B. Earth
C. Jupiter
D. Neptune

Saturn's Rings

The rings that encircle Saturn have fascinated scientists for centuries.

In November 1980, the *Voyager* mission returned pictures of Saturn's rings, providing detailed information about their composition.

Q211 What are Saturn's rings made of?

A. Liquid gas
B. Chunks of ice
C. Storm clouds
D. Granite-like rocks

A210

Which planet has the highest daily temperature?

C is the correct answer. The daily temperature of Jupiter's inner planetary layer is approximately 19,300 degrees Fahrenheit. The temperature of Jupiter's core is thought to be about 53,500 degrees Fahrenheit. Scientists remain unsure about why Jupiter radiates more than twice the amount of heat it receives from the sun.

Mercury's mean daily temperature is about 770 degrees Fahrenheit, Earth's 57 and Neptune's minus 238.

A211

What are Saturn's rings made of?

B is the correct answer. Computer-generated pictures show that Saturn's rings are actually made up of billions of small chunks of ice—some less than a yard in diameter and some as big as an office block.

Physics

An ambulance races by, and bystanders hear the frequency change of its passing siren.

Q212 What principle of physics does this audible change represent?

A. Newton's first law
B. Refraction
C. The Seventy-two rule
D. The Doppler effect

Q213 What is the scientific nomenclature for one-billionth of a second?

Q214 What is the transfer of energy or mass within a fluid owing to temperature or density differences called?

A212

What principle of physics does this audible change represent?

D is the correct answer. The Doppler effect was predicted by Austrian scientist Christian Doppler in 1842 and proved when he put an observer on one of the newly developed railroad trains. Doppler proved that the frequencies of sound waves shift according to the motion of either the source or the receiver of those waves.

Newton's first law says that matter is neither created nor destroyed.

Refraction is the deflection of light rays and energy waves as they pass from one medium into another.

The Seventy-two rule is a principle of financial accounting.

A213

What is the scientific nomenclature for one-billionth of a second?

Nanosecond

A214

What is the transfer of energy or mass within a fluid owing to temperature or density differences called?

Convection

SAMPLE GROUP TEST SCORES

Compare your score!

Now that you've completed the test and have discovered just how much—or how little—science trivia you know, you might want to see how you measure up to others around the country.

This book, like the NOVA television series, is intended for the general public, not just science enthusiasts. To that end I thought it would be fun to include the scores of a handful of groups of some more or less average types of Americans, including a number of people who earn their livings in science-related fields. So, 30 questions from this book—six from each of the five categories—were put to a rather eclectic sampling of groups ranging in size from 4 to 35, and in "type" from a 9th grade honors biology class in Massachusetts, to a group of housewives in Wisconsin, to the staff of a West Coast software company.

You can test yourself against the same questions these sample groups took. The questions given to them are listed on page 197 by their question numbers in the book (eg. Q1). Use the space to the right of each number as a scorecard as you go back to each question listed to find out if you got it right. Then compare your score with the scores of the sample groups listed on page 196.

Sample group test scores

Sample Group	Range of correct answers High / Low	Average correct	Average score
Newspaper and magazine science writers from major cities around the country	22 / 10	16	53%
Scientific and technical staff of Southwest observatory	21 / 8	16	53%
West Coast software company	19 / 12	16	53%
Writers, researchers and editors of major science magazine	20 / 8	15	50%
Public Information department of space research center in Southwest	17 / 13	15	50%
Housewives from Kenosha, Wisconsin	19 / 6	13	43%
New York stockbrokers	22 / 6	12	40%
9th grade honors biology class from public high school in Northeast	20 / 4	11	36%
Staff of microbiology and molecular genetics department of major East Coast medical school	21 / 4	10	33%
Boy's youth group from Cambridge, Massachusetts	10 / 1	6	20%
Girl's youth group from greater Boston area	6 / 0	1	3%

Sample group test questions

1. **Q1** _____
2. **Q7** _____
3. **Q11** _____
4. **Q16** _____
5. **Q24** _____
6. **Q37** _____
7. **Q41** _____
8. **Q45** _____
9. **Q47** _____
10. **Q56** _____
11. **Q58** _____
12. **Q68** _____
13. **Q79** _____
14. **Q87** _____
15. **Q92** _____
16. **Q94** _____
17. **Q99** _____
18. **Q116** _____
19. **Q123** _____
20. **Q127** _____
21. **Q130** _____
22. **Q142** _____
23. **Q147** _____
24. **Q156** _____
25. **Q164** _____
26. **Q168** _____
27. **Q176** _____
28. **Q179** _____
29. **Q182** _____
30. **Q209** _____

MENTOR Books of Special Interest

(0451)

☐ **THE HUMAN BRAIN: Its Capacities and Functions by Isaac Asimov.** A remarkable, clear investigation of how the human brain organizes and controls the total functioning of the individual. Illustrated by Anthony Ravielli. (623630—$4.95)

☐ **THE HUMAN BODY: Its Structure and Operation by Isaac Asimov.** A superbly up-to-date and informative study which also includes aids to pronunciation, and derivations of specialized terms. Drawings by Anthony Ravielli. (623584—$3.95)*

☐ **THE CHEMICALS OF LIFE by Isaac Asimov.** An investigation of the role of hormones, enzymes, protein and vitamins in the life cycle of the human body. (624181—$3.95)*

☐ **THE DOUBLE HELIX by James D. Watson.** A "behind-the-scenes" account of the work that led to the discovery of DNA. "It is a thrilling book from beginning to end—delightful, often funny, vividly observant, full of suspense and mounting tension . . . so directly candid about the brilliant and abrasive personalities and institutions involved . . ."—*New York Times.* Illustrated.
(623878—$3.50)*

☐ **INSIDE THE BRAIN by William H. Calvin, Ph.D. and George A. Ojemann, M.D.** In the guise of an operation, we are given a superbly clear, beautifully illustrated guided tour of all that medical science currently knows about the brain—including fascinating new findings. (623975—$4.50)*

*Prices higher in Canada.

Buy them at your local bookstore or use this convenient coupon for ordering.

NEW AMERICAN LIBRARY
P.O. Box 999, Bergenfield, New Jersey 07621

Please send me the books I have checked above. I am enclosing $_____(please add $1.00 to this order to cover postage and handling). Send check or money order—no cash or C.O.D.'s. Prices and numbers are subject to change without notice.

Name_____

Address_____

City_____State_____Zip Code_____

Allow 4-6 weeks for delivery.
This offer is subject to withdrawal without notice.

Scientific Reading from PLUME and MERIDIAN

(0452)

- ☐ **FROZEN STAR Of Pulsars, Black Holes, and the Fate of Stars by George Greenstein.** "This engrossing book recounts the exciting history of research on pulsars and black holes with a skillful and very entertaining blend of scientific discussion and anecdotes . . . a joy to read."—*Library Journal* Winner of the 1984 American Institute of Physics/U.S. Steel Foundation Science Writing Award and the Phi Beta Kappa Science Award for 1984. Illustrated.
(256933—$8.95)

- ☐ **THE BLUE PLANET by Louise Young.** A wonderful combination of beautiful prose and fascinating detail. *The Blue Planet* explores the latest developments in geology and earth science and shows how they have led to an exciting new view of the earth. Winner of the 1983 Carl Sandburg Award. "Popular science writing in a high tradition."—*Boston Globe*. Maps. (007089—$8.95)

- ☐ **OF A FIRE ON THE MOON by Norman Mailer.** Marked by the wit, the penetrating insight, and the philosophical scope which have long distinguished his work, Norman Mailer has written the story of the moon landing and exposes the heroic, and sinister aspects of the science of space. "A magnificent book . . . infinitely rich and complex."—*New York Times* (253772—$7.95)

- ☐ **NEW EARTHS: Restructuring Earth and Other Planets by James Edward Oberg.** Drawing upon data gathered during decent space flights, the author—mission flight controller at the NASA Johnson Space Center—examines the challenge of restructuring other planets. Combining scientific fact with fictional vignettes, here is a unique glimpse of what our future may hold on earth—and in space.
(006236—$8.95)

- ☐ **SPACEWAR by David Ritchie.** A fascinating, readable account of the militarization of outer space that examines the technology—from the earliest experiments in rocketry to the latest developments in laser and particle beam weaponry—and the political maneuvers that have merged space programs with defense departments in both the U.S. and Soviet Union. (254612—$6.95)

All prices higher in Canada.

To order, use the convenient coupon on the next page.

More Science from MERIDIAN and PLUME

(0452)

☐ **MISSION TO MARS: Plans and Concepts for the First Manned Landing by James Edward Oberg.** Every aspect of the proposed Mars program—from basic spaceship design, propulsion, navigation, and life support systems to inflight experiments, Martian exploration, Russian competition, planetary colonization, and terraforming is explored by the author, mission flight controller at the NASA Johnson Space Center. Incorporating what we know about Mars with technologies currently being developed, here is an exciting, realistic preview of the most exciting challenge in the coming decades—the first manned landing on another world.   (006554—$6.95)

☐ **ALBERT EINSTEIN: CREATOR AND REBEL by Banesh Hoffman with the collaboration of Helen Dukas.** On these pages we come to know Albert Einstein, the "backward" child, the academic outcast, the reluctant world celebrity, the exile, the pacifist, the philosopher, the humanitarian, the tragically saddened "father" of the atomic bomb, and above all, the unceasing searcher after scientific truth.   (257034—$6.95)

☐ **THE IMAGE OF ETERNITY: Roots of Time in the Physical World by David Park.** "Provocative thought about such things as the direction of time, the difference between 'time' as viewed in mechanics and 'time' in thermodynamics, and past, present and future . . . discussed with due regard to the total sweep from Pre-Aristotelean ideas through Newtonian notions to relativity and curvature of space."—*The Physics Teacher* Winner of the 1980 Phi Beta Kappa Science Award. Illustrated.   (005515—$5.95)

☐ **PHYSICS AS METAPHOR by Roger Jones.** A mind-expanding exploration of the human side of science. The author challenges the hallowed canon of science—that scientists can be truly objective about the physical universe—exposing the insubstantiality of the most "solid truths" of physics.   (007216—$7.95)

Prices slighly higher in Canada.

Buy them at your local bookstore or use this convenient coupon for ordering.

NEW AMERICAN LIBRARY, INC.
P.O. Box 999, Bergenfield, New Jersey 07621

Please send me the PLUME and MERIDIAN BOOKS I have checked above. I am enclosing $_____(please add $1.50 to this order to cover postage and handling). Send check or money order—no cash or C.O.D.'s. Prices and numbers are subject to change without notice.

Name_____

Address_____

City_____State_____Zip Code_____

Allow 4-6 weeks for delivery.
This offer is subject to withdrawal without notice.